SpringerBriefs in Energy

SpringerBriefs in Energy presents concise summaries of cutting-edge research and practical applications in all aspects of Energy. Featuring compact volumes of 50 to 125 pages, the series covers a range of content from professional to academic. Typical topics might include:

- A snapshot of a hot or emerging topic
- A contextual literature review
- A timely report of state-of-the art analytical techniques
- An in-depth case study
- A presentation of core concepts that students must understand in order to make independent contributions.

Briefs allow authors to present their ideas and readers to absorb them with minimal time investment.

Briefs will be published as part of Springer's eBook collection, with millions of users worldwide. In addition, Briefs will be available for individual print and electronic purchase. Briefs are characterized by fast, global electronic dissemination, standard publishing contracts, easy-to-use manuscript preparation and formatting guidelines, and expedited production schedules. We aim for publication 8–12 weeks after acceptance.

Both solicited and unsolicited manuscripts are considered for publication in this series. Briefs can also arise from the scale up of a planned chapter. Instead of simply contributing to an edited volume, the author gets an authored book with the space necessary to provide more data, fundamentals and background on the subject, methodology, future outlook, etc.

SpringerBriefs in Energy contains a distinct subseries focusing on Energy Analysis and edited by Charles Hall, State University of New York. Books for this subseries will emphasize quantitative accounting of energy use and availability, including the potential and limitations of new technologies in terms of energy returned on energy invested. The second distinct subseries connected to SpringerBriefs in Energy, entitled Computational Modeling of Energy Systems, is edited by Thomas Nagel, and Haibing Shao, Helmholtz Centre for Environmental Research - UFZ, Leipzig, Germany. This sub-series publishes titles focusing on the role that computer-aided engineering (CAE) plays in advancing various engineering sectors, particularly in the context of transforming energy systems towards renewable sources, decentralized landscapes, and smart grids.

All Springer brief titles should undergo standard single-blind peer-review to ensure high scientific quality by at least two experts in the field.

Thomas I. Strasser · Mihai Calin ·
Leonard Enrique Ramos Perez
Editors

European Guide to Smart Energy System Testing

The ERIGrid 2.0 Approach for Evaluating Complex Smart Energy System Configurations

Editors
Thomas I. Strasser
Center for Energy
AIT Austrian Institute of Technology
Vienna, Austria

Mihai Calin
Center for Energy
AIT Austrian Institute of Technology
Vienna, Austria

Leonard Enrique Ramos Perez
European Distributed Energy Resources
Laboratories (DERlab) e.V.
Kassel, Germany

ISSN 2191-5520   ISSN 2191-5539 (electronic)
SpringerBriefs in Energy
ISBN 978-3-031-99450-0   ISBN 978-3-031-99451-7 (eBook)
https://doi.org/10.1007/978-3-031-99451-7

This work was supported by Horizon 2020 Framework Programme (870620).

© The Editor(s) (if applicable) and The Author(s) 2025. This book is an open access publication.

Open Access This book is licensed under the terms of the Creative Commons Attribution 4.0 International License (http://creativecommons.org/licenses/by/4.0/), which permits use, sharing, adaptation, distribution and reproduction in any medium or format, as long as you give appropriate credit to the original author(s) and the source, provide a link to the Creative Commons license and indicate if changes were made.
The images or other third party material in this book are included in the book's Creative Commons license, unless indicated otherwise in a credit line to the material. If material is not included in the book's Creative Commons license and your intended use is not permitted by statutory regulation or exceeds the permitted use, you will need to obtain permission directly from the copyright holder.
The use of general descriptive names, registered names, trademarks, service marks, etc. in this publication does not imply, even in the absence of a specific statement, that such names are exempt from the relevant protective laws and regulations and therefore free for general use.
The publisher, the authors and the editors are safe to assume that the advice and information in this book are believed to be true and accurate at the date of publication. Neither the publisher nor the authors or the editors give a warranty, expressed or implied, with respect to the material contained herein or for any errors or omissions that may have been made. The publisher remains neutral with regard to jurisdictional claims in published maps and institutional affiliations.

This Springer imprint is published by the registered company Springer Nature Switzerland AG
The registered company address is: Gewerbestrasse 11, 6330 Cham, Switzerland

If disposing of this product, please recycle the paper.

Preface

Future Energy Systems—A Complex System of Systems

The drive towards a sustainable energy supply in Europe is based on integrating distributed, renewable energy resources such as photovoltaic, wind, and other sources. The stochastic nature of these generation methods presents energy utilities and network operators with the challenge of managing a more complex energy supply infrastructure. Furthermore, advancements in technology have introduced partly controllable loads, including energy storage systems, heat pumps, electric vehicle supply equipment, and other components. This, coupled with the integration of various energy sources, evolving regulatory frameworks, and market liberalization, necessitates adaptations in system operations. Sophisticated design approaches, combined with effective operational concepts and intelligent automation, are essential to transform the existing infrastructure into an intelligent entity, commonly referred to as "Smart Grids," "Smart Energy Systems," or "Cyber-Physical Energy Systems."

The increasing complexity of energy systems is evident in the enhanced coupling across domains such as electricity, automation and control, and information technology, as well as in the scale and heterogeneity of these systems. This complexity is driven by the vision of smart energy networks, which adopt a Systems of Systems (SoS) approach, integrating critical infrastructure systems and multi-energy vector systems (energy and data/information). The bidirectional flow of energy and data/information, along with the large-scale integration of renewable sources and controllable loads, adds to the complexity of planning and operating the energy infrastructure. Significant trends include the digitalization of energy networks, deeper consumer involvement, and increased market interaction, alongside the interlinking of electricity, gas, and heat grids to enhance flexibility and resilience.

Over the past two decades, numerous research and technology developments, as well as innovation activities, have been conducted under European framework programs. These efforts aim to meet the goals of the European Commission's Strategic Energy Technology Plan, called SET-Plan, for a sustainable environment and to promote innovation in cost-effective low-carbon technologies. Various demonstration projects have shown the applicability of smart grid approaches, and the concept has been extended to multi-domain energy systems. However, system-level testing is expected to play a larger role in developing future solutions, as current validation approaches and tools for smart grids and smart energy systems are not yet mature. A key element in this context is the need for educated professionals who understand the methods for validating complex smart grids and smart energy systems in a multi-domain and cyber-physical manner, which is currently lacking.

Who Should Read This Book

Before deployment, smart energy systems must be validated and tested. Industry and researchers have recognized this challenge, leading to numerous global projects. However, current validation focuses mainly on the device level, simplifying other components to electrical equivalents. This traditional approach raises questions about the integrated system's global behavior. Effective communication among specialists and the integration of different technologies are essential. Validation and test facilities often specialize in specific areas, limiting their ability to validate entire systems. To support the development of smart energy systems, system-level tests across all relevant domains and areas are needed. Alternative approaches, such as virtual or semi-virtual experiments, are being explored, but their real-world applicability remains uncertain.

This book provides an overview of the achievements and results from the "European Research Infrastructure Supporting Smart Grid and Smart Energy Systems Research, Technology Development, Validation, and Roll Out—Second Edition" (ERIGrid 2.0) project. European research and technology infrastructures play a crucial role in providing resources, conducting research, fostering innovation, and acting as knowledge hubs. However, only a few of them focus on power systems or smart energy solutions as ERIGrid 2.0 does. This book, therefore, targets professionals, engineers, researchers, young students, educators, and policymakers involved in the development and validation of new applications, solutions, and technologies related to smart energy systems.

Contribution

This book presents the major outcomes of the ERIGrid 2.0 project, funded by the European Commission under Grant Agreement No. 870620. It focuses on smart energy systems validation, testing methods, concepts, and tools, resulting from 5 years of collaboration among 20 partners across 13 European countries.

Building on the foundation laid by its predecessor, ERIGrid, which emphasized power systems and information and communication technology approaches (summarized in the "European Guide to Power System Testing"), this book can be read independently. The following eleven chapters delve into smart energy system validation approaches, concepts, tools, and examples. Chapter "Towards Energy System Validation" outlines the need for system-level validation relevant to all stakeholders. Chapters "Holistic Smart Energy System Validation" and "Enhanced Validation Methods and Benchmark" introduce validation methods and benchmarks for researchers and academics. Chapters "Extended Co-Simulation Approaches"–"Laboratory Infrastructure Integration and Automation" focus on co-simulation, real-time simulation, Hardware-in-the-Loop (HIL), and infrastructure coupling for RI/TI/laboratory operators. Chapters "Sector Coupling and Multi-Domain Systems Validation" and "Experiences with Smart System Integration and Validation" present experiences with developed methods for industry professionals. Chapter "Education Needs, Methods and Tools" addresses educators with training methods and tools. Chapter "Standardisation, Policies and Interoperability" provides suggestions and recommendations for policymakers and standardization bodies. Finally, Chapter "Summary and Future Directions" offers conclusions and future needs for all stakeholders.

Project results are available on the project website,[1] the related Zenodo Community[2] and GitHub Repository[3] as well as the funding agency's website.[4] The book's content is based on findings from relevant project deliverables and publications.

Vienna, Austria
Vienna, Austria
Kassel, Germany
April 2025

Thomas I. Strasser
Mihai Calin
Leonard Enrique Ramos Perez

---

[1] https://erigrid2.eu.
[2] https://zenodo.org/communities/erigrid2.
[3] https://github.com/ERIGrid2.
[4] https://cordis.europa.eu/project/id/870620.

Acknowledgments

We express our gratitude to all ERIGrid 2.0 partners who contributed to the development of this book, especially E. Rikos (CRES), J. Karnsamrong (OFFIS), J. S. Schwarz (OFFIS), Z. Feng (UoS), A. Acosta (RWTH), E. Widl (AIT), G. Silano (RSE), A. Kontou, and P. Kotsampopoulos (ICCS-NTUA) for coordinating the writing of the respective chapters. Special thanks to E. Mrakotsky-Kolm (AIT) for reviewing the entire book. We also extend our appreciation to S. K. Mathiyazhagan and A. Doyle from Springer for their invaluable assistance during the writing and editing process. Lastly, we are grateful to the European Commission for their financial support, which made this work possible.

Vienna, Austria Thomas I. Strasser
Vienna, Austria Mihai Calin
Kassel, Germany Leonard Enrique Ramos Perez
April 2025

Contents

Towards Energy System Validation 1
T. I. Strasser, M. Calin, and L. E. Ramos Perez

Holistic Smart Energy System Validation 9
F. Pröstl Andren, E. Widl, K. Heussen, E. Rikos, T. I. Strasser, and P. Raussi

Enhanced Validation Methods and Benchmark 23
J. Kamsamrong, E. Widl, J. S. Schwarz, K. Heussen, P. Raussi, G. Arnold, A. De Paola, Z. Feng, O. Werth, M. C. Pham, and Q. T. Tran

Extended Co-Simulation Approaches 35
J. S. Schwarz, E. Widl, O. Gehrke, K. Heussen, R. Fabian, and J. Kamsamrong

Improved Hardware-in-the-Loop-Based Testing 47
Z. Feng, M. H. Syed, A. Paspatis, A. Kontou, G. Lauss, A. De Paola, P. Kotsampopoulos, N. Hatziargyriou, and G. Burt

Laboratory Infrastructure Integration and Automation 65
A. Acosta, G. Silano, G. Paludetto, O. Gehrke, M. C. Pham, Q. T. Tran, V. Rajkumar, S. Vogel, and A. Monti

Sector Coupling and Multi-Domain Systems Validation 77
E. Widl, G. Silano, O. Gehrke, and T. Zerihun

Experiences with Smart System Integration and Validation 87
G. Silano, A. Kontou, Z. Feng, A. Acosta, O. Gehrke, T. Zerihun, S. Sanchez-Acevedo, G. Arnold, J. E. Rodriguez-Seco, F. Lazzari, C. Rodio, and L. Pellegrino

Education Needs, Methods and Tools 105
A. Kontou, P. Kotsampopoulos, K. Heussen, L. E. Ramos Perez, T. I. Strasser, E. Rikos, G. Makrides, Z. Feng, P. Karafotis, A. Paspatis, M. Syed, G. Lauss, M. Calin, and N. Hatziargyriou

Standardisation, Policies and Interoperability 119
M. Calin, G. Lauss, L. E. Ramos Perez, and T. I. Strasser

Summary and Future Directions 129
T. I. Strasser, M. Calin, and L. E. Ramos Perez

Acronyms

AC	Alternating Current
ADC	Analog-to-Digital Conversion
AI	Artificial Intelligence
API	Application Programming Interface
BESS	Battery Energy Storage System
CD	Continuous Deployment
CHIL	Controller Hardware-in-the-Loop
CHP	Combined Heat and Power
CI	Continuous Integration
CI/CD	Continuous Integration and Continuous Delivery/Deployment
CM	Configuration Management
CPES	Cyber-Physical Energy System
CVC	Coordinated Voltage Control
DAC	Digital-to-Analog Conversion
DC	Direct Current
DER	Distributed Energy Resources
DG	Distributed Generation
DN	Distribution Network
DoE	Design of Experiments
DPSL	Dynamic Power Systems Laboratory
DRL	Deep Reinforcement Learning
DRTS	Digital Real-Time Simulator
DSO	Distribution System Operator
DT	Digital Twin
DuI	Domain Under Investigation
EESL	Electric Energy System Laboratory
EMP	Environmental Measurements Platform
EMT	Electromagnetic Transients
ERA	European Research Area
ERIC	European Research Infrastructure Consortium
ES	Experiment Specification

ESFRI	European Strategy Forum on Research Infrastructures
EU	European Union
FAIR	Findable, Accessible, Interoperable and Reusable
FBF	Feedback Filtering
FMI	Functional Mockup Interface
FMU	Functional Mockup Unit
FRT	Fault-Ride-Through
GD-PHIL	Geographically Distributed Power Hardware in the Loop
GDRTS	Geographically Distributed Real-Time Simulation
GDS	Geographically Distributed Simulation
GFC	Grid-Forming Converter
GFL	Grid-Following
GFM	Grid-Forming
GM	Gain Margin
GRI	Global Research Infrastructure
GUI	Graphical User Interface
HIL	Hardware-in-the-Loop
HOI	Hardware of Interest
HTD	Holistic Test Description
I/O	Input and Output
IA	Interface Algorithm
IasC	Infrastructure as Code
ICT	Information and Communication Technology
IEC	International Electrotechnical Commission
IEEE	Institute of Electrical and Electronics Engineers
IoT	Internet of Things
IR	Impedance Ratio
IRENA	International Renewable Energy Agency
ISO	International Organization for Standardization
ITM	Ideal Transformer Model
KPI	Key Performance Indicator
LPF	Low-Pass Filter
LSTM	Long Short-Term Memory
LV	Low Voltage
MG	Microgrid
MOI	Model of Interest
MOOC	Massive Open Online Course
MV	Medium Voltage
OECD	Economic Co-operation and Development
OLTC	On-Load Tap Changer
OPC UA	OPC Unified Architecture
PA	Power Amplifier
PAC	Programmable Automation Controller
PCC	Point of Common Coupling
PFC	Power Flow Calculator

PHIL	Power Hardware-in-the-Loop
PI	Power Interface
PLL	Phase Locked Loop
PM	Phase Margin
PoI	Purpose of Investigation
PV	Photovoltaics
QS	Qualification Strategy
R&D	Research and Development
RES	Renewable Energy Source
RI	Research Infrastructure
RIasC	Research Infrastructure as Code
RMS	Root Mean Square
RTI	Research and Technology Infrastructure
RTS	Real-Time Simulation or Real-Time Simulator
SA	Sensitivity Analysis
SC	System Configuration
SCADA	Supervisory Control and Data Acquisition
SDFT	Sliding Discrete Fourier Transform
SIL	Software-in-the-Loop
SNR	Signal-to-Noise
SoI	System of Interest
SoS	Systems of Systems
SP	Smith Predictor
SS	Smart System
SuT	System under Test
TA	Transnational Access
TC	Test Case
TCP	Test Case Profile
THD	Total Harmonic Distortion
THD+N	Total Harmonic Distortion plus Noise
TI	Technology Infrastructure
TS	Test Specification
TSO	Transmission System Operator
uAPI	Universal API
UML	Unified Model Language
UP	User Project
UQ	Uncertainty Quantification
USAT	Uncertainty Structure Analysis Tool
VSI	Virtual Shifting Impedance
VSIDC	Virtual Shifting Impedance with Delay Compensation
VSM	Virtual Synchronous Machine

Towards Energy System Validation

T. I. Strasser, M. Calin, and L. E. Ramos Perez

Abstract The evolution of interconnected multi-energy networks has led to significant advancements, integrating various energy carriers like electricity, thermal, gas, heating, and cooling. This integration offers benefits such as increased flexibility and low-carbon technologies but also introduces challenges due to the variability of renewable energy sources and the complexity of managing distributed energy resources. The ERIGrid 2.0 approach addresses these challenges by providing a comprehensive, cyber-physical systems-oriented framework for validating smart energy systems, facilitating research and innovation through advanced validation methods and tools.

1 Higher Complexity in Multi-Energy Systems

Energy systems have evolved to a point in which multiple systems, which traditionally operated independently, are now interconnected, functioning as a single macrosystem. Such multi-energy systems combine different energy carriers, including electricity, thermal, gas, heating, and cooling as an integrated network both spatially and operatively. Under this concept, the energy carriers' infrastructure alongside the different stages in the process (generation, transmission, distribution and commercialisation) are interconnected in a complex infrastructure by exploiting the link between them [7]. This transformation of energy systems led to numerous developments in terms of technologies, tools and methodologies for the integration of different energy carriers and dynamics of the energy markets [1].

T. I. Strasser (✉) · M. Calin
AIT Austrian Institute of Technology, Vienna, Austria
e-mail: thomas.strasser@ait.ac.at

M. Calin
e-mail: mihai.calin@ait.ac.at

L. E. Ramos Perez
European Distributed Energy Resources Laboratories (DERlab) e.V., Kassel, Germany
e-mail: leonard.ramos@der-lab.net

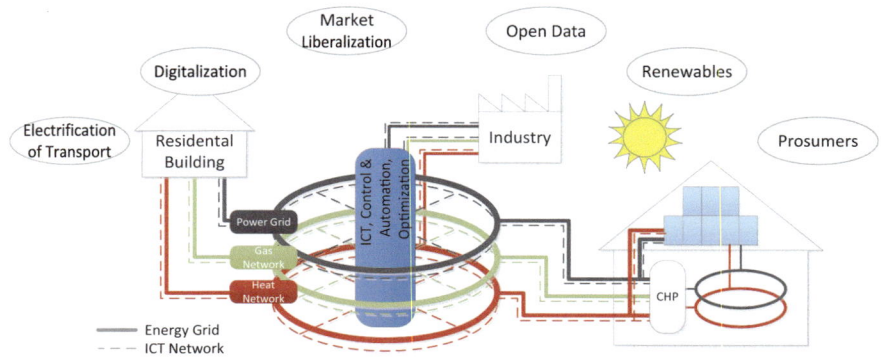

Fig. 1 Example of a cyber-physical energy system

Although the implementation of multi-energy systems shows several advantages, such as higher flexibility, the use of low-carbon technologies, and the decentralisation of generation, in parallel, other challenges are arising. For instance, due to the introduction of a high share of Renewable Energy Sources (RES), the variability in generation increases and causes more complexity due to the uncertain nature of the primary energy sources. In addition, the implementation of Distributed Energy Resource (DER) generation and interconnection of multiple energy carriers make it more challenging to manage and optimise the system operation [2]. In addition to that, the control of such a system needs advanced Information and Communication Technology (ICT), resulting in a so-called Cyber-Physical Energy System as shown in Fig. 1. Thus, the modelling of these systems becomes so complex due to the robustness and both number and variety of assets involved that in most cases, simplifications and initial assumptions are made, which may lead to inadequate operation configuration and optimization [7].

2 Existing Validation Approaches and Research Needs

Current validation approaches in the power and energy systems domain often focus on individual domains, sub-domains, or specific components [9–11]. There is a pressing need for more integrated and comprehensive validation frameworks that address the interdependencies and dynamic interactions within multi-energy systems. Typical validation cycles involve creating simulation models to estimate real-world infrastructure behaviour, which are then validated through hardware experiments.

Simplification of models and real-world fluctuations lead to discrepancies between simulation models and hardware experiments, resulting in uncertainty. Traditional probabilistic or boundary load flow calculations do not fully address the broader range of fluctuations and uncertainties in cyber-physical energy systems. Moreover, in co-simulation-based approaches, synchronization errors can occur due to different time management methods, and there is limited integration of different power simulation tools, which is necessary for complex power system analysis. Real-time simulation and Hardware-in-the-Loop (HIL) technologies often neglect interactions between hardware components and other units due to laboratory limitations and test setup simplifications. Transient and dynamic characteristics may not be considered due to the lack of interaction in the applied setup. Furthermore, issues with the initialization of large distributed or remote Power Hardware-in-the-Loop (PHIL) setups need to be addressed for wider use [10].

Additionally, the integration, coupling, and automation of RIs and testing facilities have shown promising advancements. Virtual integration of hardware and software assets located at geographically dispersed locations has been investigated, demonstrating the feasibility and benefits of real-time virtual connections for joint operation in single experiments [9]. However, harmonized approaches with a universal access interface are still missing.

Therefore, further research is required in areas such as the development of standardized validation procedures, the integration of real-time data into validation processes, and the creation of advanced co-simulation and HIL-based platforms capable of handling the complexity of these systems as depicted in Fig. 2.

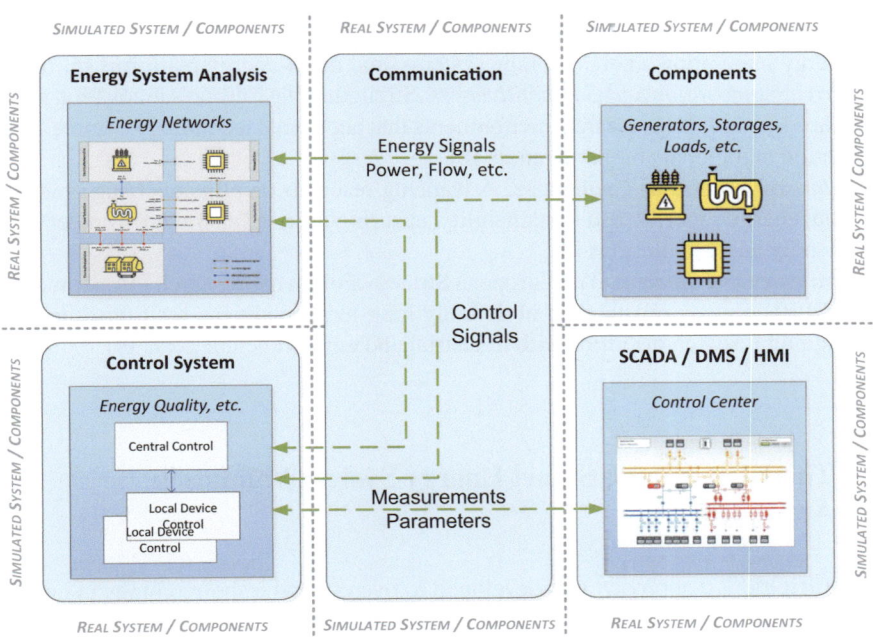

Fig. 2 CPES validation by using virtual and real components (adapted from [9])

3 The Role of Validation and Testing Facilities

A Research Infrastructure (RI) can be defined as the conglomeration of physical and human resources built up for research, whereas a Technology Infrastructure (TI) focuses more on development activities towards market adoption. The definition does not only apply to physical facilities like laboratories, which are common in the power and energy domain, but also to Information and Communication Technology (ICT), storage databases, instrumentation, and others.

In Europe, RIs, TIs and testing facilities serve as key drivers of scientific progress, innovation, and industrial competitiveness, particularly in power and energy systems. Their core roles include:

- *Enabling Cutting-Edge Research:* Providing researchers and industrial stakeholders with access to state-of-the-art equipment, data, and expertise, fostering scientific breakthroughs beyond individual institutional capabilities [4].
- *Facilitating Collaboration:* Acting as national and international partnerships hubs, connecting experts from academia, industry, and policymaking institutions. Initiatives like ERIGrid and its successor, ERIGrid 2.0, focus on unifying European smart grid and smart energy systems RIs/TIs and expanding access for external research groups.
- *Supporting Technological Development and Validation:* This is essential for testing and validating emerging energy technologies, including smart grids, renewable energy integration, energy storage systems, and multi-energy platforms [5, 6].
- *Driving Innovation and Competitiveness:* Strengthening Europe's innovation landscape by providing research environments that accelerate technology commercialization and global competitiveness [8].
- *Addressing Societal Challenges:* Advancing research on climate change mitigation, energy security, and sustainability, ensuring more efficient, environmentally friendly energy solutions [3].
- *Strategic Development:* The European Strategy Forum on Research Infrastructures (ESFRI) plays a pivotal role in defining long-term roadmaps for European RIs, aligning research priorities with industrial and environmental goals [4].

4 The ERIGrid 2.0 Smart Energy System Validation Approach

The ERIGrid 2.0 approach was developed to overcome the shortcomings in energy infrastructure evaluation. It addresses the open points by conceiving a holistic, cyber-physical systems-oriented approach for the testing of smart energy systems. This integrated European smart energy systems Research and Technology Infrastructure (RTI) targets the following points:

- Provide a comprehensive framework for the validation of smart energy systems, considering their cyber and physical aspects.
- Offer access to a wide range of interconnected validation and testing facilities across Europe.
- Develop and disseminate advanced validation methods, tools, and best practices for the energy research community.

Thus, ERIGrid 2.0 plays a pivotal role in facilitating research and innovation in the domain of power and energy systems. By providing access to a broad spectrum of advanced services, methods, and tools, the initiative enables researchers to conduct a comprehensive validation of complex smart energy solutions. A core part of the project is the holistic, cyber-physical validation of energy infrastructures, ensuring that both the digital and physical aspects of energy networks are assessed with high fidelity. Additionally, the project emphasizes virtual testing, allowing researchers to leverage remote access to validation tools without requiring physical presence at testing facilities. Therefore, ERIGrid 2.0 integrates cutting-edge methodologies designed to enhance testing capabilities for smart energy systems. While a fully exhaustive list of developed validation tools is not readily available, key aspects and methodologies have been widely documented, forming the foundation of its comprehensive approach. The following points highlight the core components of ERIGrid 2.0's validation strategies:

- *Holistic and Cyber-Physical Systems-Based Validation:* Central to ERIGrid 2.0, this approach examines the interdependencies between cyber and physical components in smart energy systems, ensuring accurate assessments across real-world operational conditions.
- *Virtual Testing:* The initiative grants access to remote facilities, services, and validation environments, reducing the barriers to testing complex energy systems.
- *Co-simulation and Hardware-in-the-Loop (HIL)-based Testing:* By coupling domain-specific simulation tools and also by integrating physical hardware with simulated models, researchers can evaluate system behaviour under a variety of operational scenarios. ERIGrid 2.0 also refines Controller Hardware-in-the-Loop (CHIL) and Power Hardware-in-the-Loop (PHIL) methodologies to enhance their stability and effectiveness.
- *Benchmark Models:* ERIGrid 2.0 contributes to the development of standardized, interoperable benchmark models, enabling consistent testing and comparative analysis of results.
- *Real-Time Coupling of Geographically Distributed Research Infrastructures:* Methods were explored to interconnect remote testing facilities, enabling coordinated validation scenarios across multiple locations.

Training and education concepts are also provided to support the overall research and technology development activities for advanced validation and testing. An interesting point in the ERIGrid 2.0 approach is to provide free access to the integrated

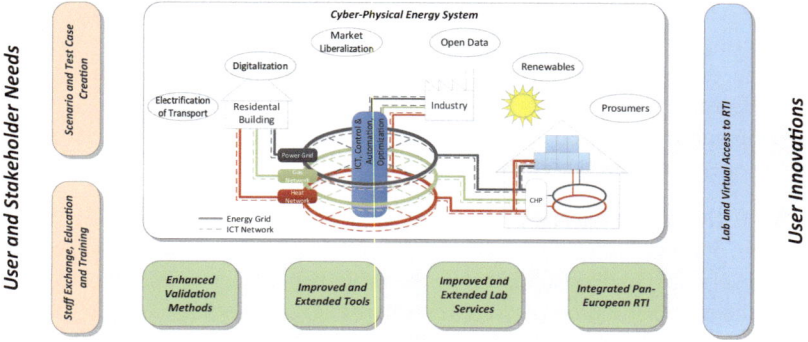

Fig. 3 Overview of the ERIGrid 2.0 approach

RTIs (i.e., partner's smart grid and energy system laboratories) in a physical but also virtual manner and the corresponding methods and tools for external user groups from industry and academia as outlined in Fig. 3.

References

1. Chertkov M, Andersson G (2020) Multienergy systems. Proc IEEE 108(9):1387–1391. https://doi.org/10.1109/JPROC.2020.3015320
2. Chicco G, Riaz S, Mazza A, Mancarella P (2020) Flexibility from distributed multienergy systems. Proc IEEE 108(9):1496–1517. https://doi.org/10.1109/JPROC.2020.2986378
3. European Commission: Energy Research and Innovation Strategy (2020). https://ec.europa.eu/energy/home_en
4. European Strategy Forum on Research Infrastructures (ESFRI): European Strategy Forum on Research Infrastructures Roadmap. Technical report, ESFRI (2021). https://www.esfri.eu/
5. Institute of Electrical and Electronics Engineers (IEEE): Power Systems and Smart Grid Testing Standards (2022). https://standards.ieee.org/
6. International Electrotechnical Commission (IEC): Standardization for Smart Grids and Energy Systems (2019)
7. Manco G, Tesio U, Guelpa E, Verda V (2024) A review on multi energy systems modelling and optimization. Appl Therm Eng 236:121871. https://doi.org/10.1016/j.applthermaleng.2023.121871
8. Organisation for Economic Co-operation and Development: Optimising the Operation and Use of National Research Infrastructures. OECD Publishing, Paris, France (2017). https://www.oecd.org/en/publications/optimising-the-operation-and-use-of-national-research-infrastructures_7cc876f7-en.html
9. Strasser T, de Jong E, Sosnina M (2020) European guide to power system testing: the ERIGrid holistic approach for evaluating complex smart grid configurations. Springer, Berlin. https://doi.org/10.1007/978-3-030-42274-5

10. Strasser T, Pröstl Andrén F, Lauss G, Bründlinger R, Brunner H, Moyo C, Seitl C, Rohjans S, Lehnhoff S, Palensky P et al (2017) Towards holistic power distribution system validation and testing-an overview and discussion of different possibilities. e & i Elektrotechnik und Informationstechnik 134:71–77
11. Strasser TI, Andren FP, Widl E et al (2018) An integrated pan-European research infrastructure for validating smart grid systems. e & i Elektrotechnik und Informationstechnik 135(8): 616–622. https://doi.org/10.1007/s00502-018-0667-7

Open Access This chapter is licensed under the terms of the Creative Commons Attribution 4.0 International License (http://creativecommons.org/licenses/by/4.0/), which permits use, sharing, adaptation, distribution and reproduction in any medium or format, as long as you give appropriate credit to the original author(s) and the source, provide a link to the Creative Commons license and indicate if changes were made.

The images or other third party material in this chapter are included in the chapter's Creative Commons license, unless indicated otherwise in a credit line to the material. If material is not included in the chapter's Creative Commons license and your intended use is not permitted by statutory regulation or exceeds the permitted use, you will need to obtain permission directly from the copyright holder.

Holistic Smart Energy System Validation

F. Pröstl Andren, E. Widl, K. Heussen, E. Rikos, T. I. Strasser, and P. Raussi

Abstract The complexity of smart energy systems poses a significant challenge in terms of testing. As a solution to tackle the specific challenge, the Holistic Validation approach proposes the use of a template-based methodology called Holistic Test Description. The most important extensions in the methodology are presented in this work. These extensions regard the representation of Systems, Control Functions Uncertainty, and Profiling, while automation of the templates via machine-readable approaches is also examined.

1 Overview of Methodology and Motivation for Extensions

To address the issue of testing modern power and energy systems, the "Holistic Smart Energy Systems Validation" is a highly efficient systematic method. Originally developed in the predecessor ERIGrid project [1, 13] the proposed methodology,

---

F. Pröstl Andren · E. Widl · T. I. Strasser
AIT Austrian Institute of Technology, Vienna, Austria
e-mail: filip.proestl-andren@ait.ac.at

E. Widl
e-mail: edmund.widl@ait.ac.at

T. I. Strasser
e-mail: thomas.strasser@ait.ac.at

K. Heussen
Technical University of Denmark, Kgs. Lyngby, Denmark
e-mail: kheu@dtu.dk

E. Rikos (✉)
CRES Centre for Renewable Energy Sources and Saving, Pikermi, Greece
e-mail: vrikos@cres.gr

P. Raussi
VTT Technical Research Centre of Finland, Espoo, Finland
e-mail: petra.raussi@vtt.fi

also known as Holistic Test Description (HTD), provides a holistic framework for describing tests in CPES.

Modern-day power and energy systems are incorporating large shares of RES, particularly Photovolatics (PV) and wind generators, especially at the distribution network level. These technologies, alongside other DER and distributed generation, result in several challenges in the operation of power systems, including reverse power flows, congestions, etc. To address such issues, distribution networks are being modernized, incorporating technologies such as ICT for intelligent monitoring and control, as well as coupling with various energy sectors to enable better utilization, storage and management of RES energy. Due to the multiple sectors incorporated, the latter are often referred to as CPES. The development of CPES applications usually relies on four steps: (i) Design, (ii) Implementation, (iii) Validation, and (iv) Deployment (which enable an application optimization).

Concerning the Validation step, the "Holistic Smart Energy Systems Validation" developed in ERIGrid bridges the gap existing in present-day testing procedures that address the testing of CPES applications mostly at the device level. In this respect, the HTD method tackles the problem in a cyber-physical and multi-domain manner by considering the CPES complexity in its entirety. Inspired by related approaches used in the automotive industry, consumer electronics, etc., the HTD is an approach that, unlike existing methods, addresses the problem of test description in a more holistic manner and at higher complexity levels. In a nutshell, the HTD method consists of three (3) key levels: (i) Test Case (TC), (ii) Test Specification (TS), and (iii) Experiment Specification (ES). The three levels are elaborated with the use of a set of templates which, in addition to the particular templates for the three levels, include the Qualification Strategy (QS), the Experiment Realization Plan, the Results Annotation, the Experiment Evaluation, and the Test Case Canvas which provides an overview of the Test Case highlights.

Despite its holistic nature, the HTD approach is not a rigid framework in the sense that it can be used in a flexible and adaptable manner depending on the users' needs. From this perspective, this chapter presents some key adaptations of the HTD in terms of extensions that aim to enhance the methodology. The general motivation behind these extensions is the wide use of the HTD in different projects and applications and the realization that some requirements are met more frequently than others and can be generalized and incorporated in the templates of the methodology. In this context, as an ever-evolving approach, the HTD extensions reflect either customization of the general method to specific approaches or feedback by users towards streamlining the methodology. A case in point is the extensions based on the PreCISE approach [6, 7]. This method tackles the issue of a holistic simulation approach based on the HTD but implements specific modifications in the templates to cover several simulation aspects. Among these modifications, two of them, namely System Configuration and Control Function descriptions, are deemed as important, general-purpose extensions for the HTD, and as such, they are presented in this chapter in more detail alongside other template and organizational extensions.

2 Template Extensions

As a crucial part of feedback collection and continuous improvement of the methodology in ERIGrid 2.0, specific efforts were made to improve three particular aspects of the HTD templates:

- Extended system configuration description
- Control function description
- Uncertainty representation.

These extensions arise from the need for more emphasis on the above-mentioned aspects within the HTD approach, as the result of analysis within ERIGrid 2.0 but also during implementation of the HTD in other EU-funded projects such as SmILES that led to the development of the PreCISE approach. Even though these extensions are based on specific implementations of the HTD, it is evident that their applicability can be generalized and incorporated in the HTD templates, as these aspects have to be invariably addressed in almost any CPES-related test. More details regarding each particular extension are provided below, including information about the motivation and short examples.

2.1 Extended System Configuration Description

The system configuration in an HTD description is a static, comprehensive depiction of a technical system, detailing all key components. While dynamic attributes may be included, it excludes system usage information. Previously, the HTD template had only a brief system configuration section, making it difficult to describe advanced configurations, especially when dealing with complex energy systems. To address this, an optional extension was developed, offering a structured, dedicated template that allows to define the components of a system as well as their interactions and topological composition (see Fig. 1).

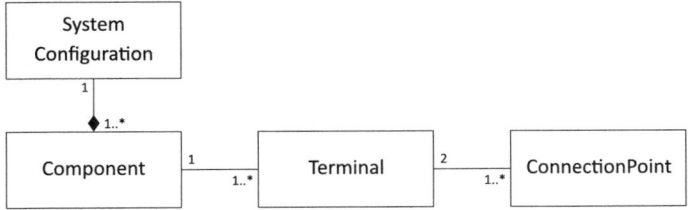

Fig. 1 Components, terminals and connection points in a system configuration

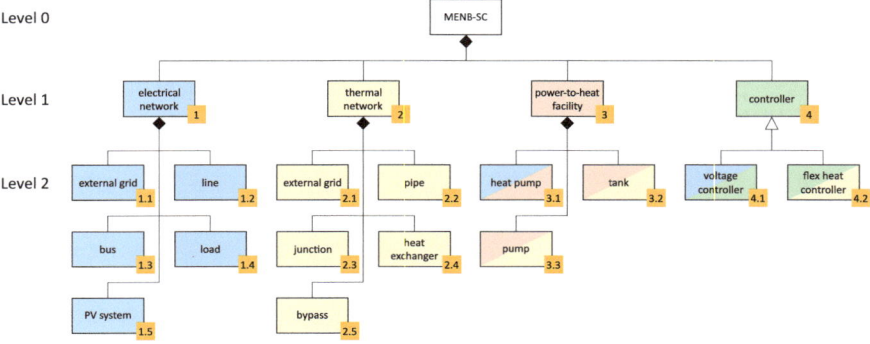

Fig. 2 Example of a system breakdown diagram

The template standardizes information exchange across stakeholders through the following items:

- General description (system name, ID, and context).
- System breakdown diagram (Unified Model Language (UML)-based overview of system elements).
- Graphical representations (domain-specific diagrams).
- Component descriptions (details on components and their attributes).
- Terminal descriptions (component interfaces).
- Connection points (terminal connections between components).

Figure 2 shows an example of a system breakdown diagram for a multi-energy network. This diagram provides a structured overview of the system configuration, breaking down the complex system into subsystems with increasing levels of detail. A complete example of a system configuration description can be found in [10].

2.2 Description of Control Functions

A control function describes an embedded control system or an aspect of it, defining its behaviour alongside the system's inherent physical properties. It formalizes how individual system components operate in a use case. Previously, the HTD template lacked a section for control functions. To address this, an optional extension was developed, offering a structured template to standardize information exchange on control function. Similar to the case of the system configuration extension, an optional control function template has been developed. This template covers these categories:

- Functional description (overview of the problem and technical background).
- Terminology (optional, definitions of key terms).

- Methodology (detailed explanation of function/algorithm, including necessary equations).
- Limitations (constraints due to input data or algorithmic properties).
- Inputs (static configuration and dynamic measurement data)
- Outputs (setpoints, events, etc.).
- Use cases (demonstrations of function requirements and applications).
- Diagrams (visual representation of data flow).
- Algorithmic functions (optional, step-by-step algorithm descriptions, including pseudocode).
- Deterministic/stochastic functions (optional, mathematical formulations).
- Deployment (optional, details on hardware/software implementation).

Control functions may be mathematical (e.g., transfer functions, stochastic processes) or algorithmic, requiring different description approaches and thus, some categories remain optional. Figure 3 shows an example of pseudocode for a control function implementing a simple voltage controller. This algorithmic description allows to implement the corresponding controller in different programming languages. A complete example of a control function description can be found in [10].

2.3 Uncertainty Representation

The need to control for uncertainty is inherent in testing. The HTD approach already reduces qualitative uncertainty by clarification of the test concept and process, which facilitates the communication between domain experts. However, quantitative uncertainty is always associated with test results in several forms, which are often not uncovered. In previous work, it had been established that HTD maps well to the Design of Experiments (DoE) methodology, however, there was no guideline to how uncertainty needs to be identified and addressed in the development of a test design. To assist test planners in controlling for quantitative uncertainty in test outcomes, the HTD has been revisited and extended with specific tools that now allow for a systematic treatment of quantitative uncertainty aspects, as illustrated in Fig. 4.

The HTD templates now offer additional guidelines for uncertainty aspects in the templates for TCs, QSs, TSs, and ESs. Further, an Uncertainty Structure Analysis Tool (USAT) was developed, which supports the structured analysis of a complete test system and experimental setup. The tool assists in the identification and systematic prioritization of uncertainty aspects in the test hypothesis, the test system configuration and the experimental setup. The USAT tool also bridges the qualitative HTD analysis to the quantitative DoE methods as well as to the DoE toolbox (cf. Sect. 3.1). Further details on the HTD uncertainty representation extensions are explained in Chap. 3 and [12].

```
BEGIN ALGORITHM (voltage controller)
    DETERMINE operation veto still pending
    IF operation veto pending THEN
        EXIT ALGORITHM (voltage controller)
    END IF

    READ meas_voltage_pu // get latest voltage measurement
    READ setpoint_hp_p_el // get latest value for setpoint

    // lower voltage band threshold depends on heat pump status
    IF heat pump is switched off THEN
        SET delta_vm_lower_pu = delta_vm_lower_pu_hp_off
    ELSE
        SET delta_vm_lower_pu = delta_vm_lower_pu_hp_on
    END IF

    COMPUTE delta_v_meas_pu = meas_voltage_pu - 1

    IF delta_vm_lower_pu < delta_v_meas_pu < delta_vm_upper_pu THEN
        IF heat pump is switched off AND NOT operation veto pending THEN
            SET setpoint_hp_p_el = hp_p_el_mw_min
            SET operation veto to pending
        END IF
    ELSE
        // update setpoint
        COMPUTE res = k_p * (delta_v_meas_pu - delta_vm_deadband) / hp_p_el_mw_step
        COMPUTE step_res = int(res)
        IF fabs(res - step_res) > hp_p_el_mw_step THEN
            SET setpoint_hp_p_el = setpoint_hp_p_el + hp_p_el_mw_step * (step_res + 1)
        END IF
        // check operational constraints
        IF new setpoint above max. heat pump consumption THEN
            SET setpoint_hp_p_el = hp_p_el_mw_rated
        ELSE IF new setpoint below min. heat pump consumption AND \
             NOT no operation veto pending THEN
            SET setpoint_hp_p_el = 0
            SET operation veto to pending
        ELSE IF new setpoint below min. heat pump consumption AND operation veto pending THEN
            SET setpoint_hp_p_el = hp_p_el_mw_min
        END IF
    END IF
END ALGORITHM (voltage controller)
```

Fig. 3 Example of pseudocode for defining a control function

3 Organisational Extensions

In addition to the template extensions presented in the previous section, efforts to improve other aspects of the HTD methodology were made in ERIGrid 2.0. These extensions refer to organizational, reporting, and communication aspects that enable a better understanding of the HTD deployment framework, especially by third parties. More detailed information is provided below regarding the three particular extensions mentioned below:

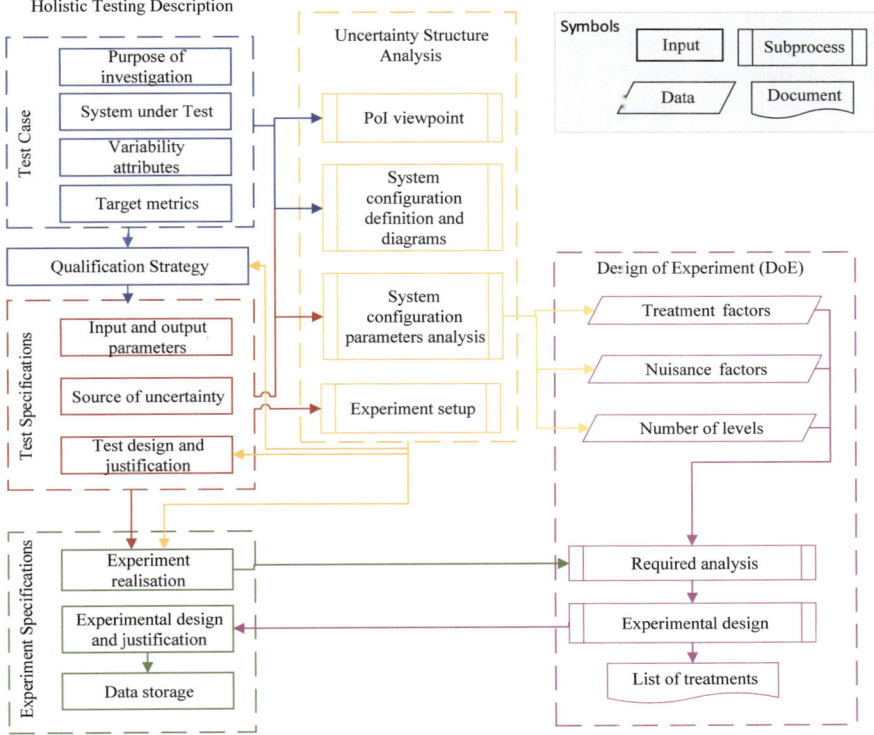

Fig. 4 Uncertainty aspects identified via HTD and USAT extension

- TC profiling.
- Automated tool to support usage of online TC repositories.
- TC repositories.

These extensions arise from the need for better organizing and publishing the large number of collected TC descriptions developed in ERIGrid 2.0 and other relevant projects. With the ever-expanding use of the HTD templates in projects and activities outside ERIGrid 2.0, the proposed extensions facilitate a better organizational context for the users. It should be noted that the term TC in this section encompasses not only the TC template but also additional elements such as the QS and the TS, which are, in short, called TC.

3.1 Recognising Test Case Profiles

The complexity of CPES results in a multitude of research, deployment, and testing areas with diverse characteristics and requirements. This diversity is also very prominent in the examples of TC descriptions developed so far. Despite the wealth of these examples, non-systematic categorization of TC may result in inefficient use of them by prospective TC developers who would like to refer to exemplary TC as guidelines to develop a particular project. The solution to this problem is the classification of diverse TC descriptions into categories with particular characteristics, called Test Case Profiles (TCPs). The latter profiles serve the purpose of holistic test procedures' harmonization, such as laboratory experiments, coupling of remote RTIs, and extensions in simulation cases. Ultimately, TCPs are equipped with exemplary TC descriptions that can be used as reference documents to further develop more targeted TCs. The harmonization of diverse TC descriptions and their classification into TCPs was tackled in ERIGrid 2.0, and the approach is presented here as good practice.

The analysis proposed in ERIGrid 2.0 consists of steps that, in an iteration approach, define specific Functional Scenarios, which are followed by the development of a detailed and exhaustive set of TCs [11]. The aforementioned scenarios stand for:

- FS1 Ancillary services are provided by DER and active grid assets.
- FS2 Microgrids and energy communities.
- FS3 Energy sector coupling.
- FS4 Frequency and voltage stability.
- FS5 Aggregation and flexibility management.
- FS6 Digitalisation.

These scenarios cover, to a large extent, a wide range of CPES applications and studies; however, for a more concise and systematic categorization of TCs, only three major technological areas are proposed:

- *Active electrical distribution grid*: Pure electrical network where several types of DERs and loads are connected.
- *Multi-domain energy systems*: Coupling of multiple energy vectors related to the smart grid environment.
- *ICT-enhanced energy systems*: Combination of multiple domains integrated in a power system, including ICT. In contrast to the first major technological area, ICT impacts the behavior of energy systems.

A second crucial aspect in the methodology is the definition of four dimensions related to the HTD descriptions and adjusted for the sake of convenience to the following nomenclature:

- Domain under Investigation (DuI).
- Phenomenon under Test.
- Type of Assessment.
- Test System/Components.

Holistic Smart Energy System Validation

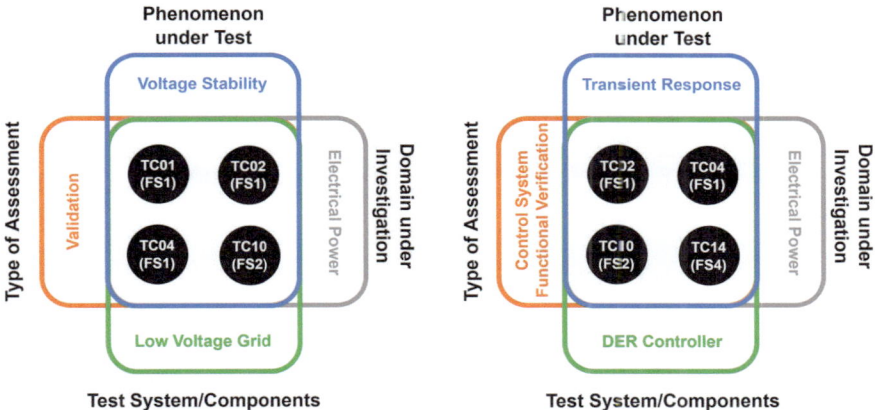

Fig. 5 Two TCP examples in the area of "Active electrical distribution grid"

For the easiest association of TC characteristics to the dimensions mentioned above, a list of keywords is defined for each dimension. For example, Ancillary Services is one of the keywords for the Phenomenon under Test, and Medium Voltage Grid for the Test System/Components. By applying the criteria mentioned above, seemingly disconnected TC descriptions can be grouped together and even graphically illustrated as shown in the example of Fig. 5, which refers to the first area (Active electrical distribution grid) and categorizes five different TC descriptions into two groups defined by the shown combinations of dimensions.

In the example of Fig. 5 the listed TCs are: *Voltage control with an on-load tap changer controller (TC01)*, *Complying with the Fault-Ride-Through (FRT) requirements in inverter-based droop-controlled microgrids* (TC02), *Investigation of different voltage control techniques for inverter-interfaced DERs in microgrids (TC04)*, *Evaluation of secure transition from grid-connected to islanded operation: uninterruptible power supply (TC10)*, and *Synthetic inertia and fast frequency response/control provided by converter-based resources (TC14)*.

3.2 Making the HTD Templates Machine Readable

One of the goals of this work was to improve the reusability of TCs. The HTD developed in the predecessor ERIGrid project [9] was based on a Word template, and it was then up to the user to publish this or make it available to other users. A disadvantage of this method is that it is difficult for other users to browse and search for information when this is only available in a file format. Therefore, the goal here was to create an online repository where the TCs can be published and browsed by other users.

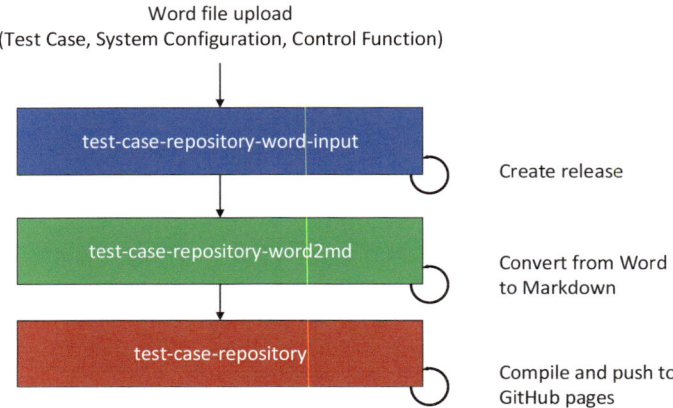

Fig. 6 Conversion of TC Word files to online repository

Different options were discussed, such as creating a completely new approach where the HTD is moved completely online. In the end, the choice fell on an approach that allows easy online publication of TCs but at the same time stays backwards compatible with the old templates. The developed solution uses GitHub,[1] GitHub Actions,[2] and GitHub Pages[3] to convert uploaded Word files into web pages. The workflow is depicted in Fig. 6.

The process starts with the upload of a file to the *test-case-repository-word-input* GitHub Repository [4]. When triggered, all compatible Word files are collected. This can either be TCs written in the original template, see Sect. 1, or templates from one of the extensions, see Sect. 2. Using the collected files, a new release is created in the repository. The release provides a link that can be used to access the collected files.

Once the release has been created, the *test-case-repository-word2md* repository is automatically triggered [5]. When a trigger is received, this repository downloads the Word files from the release in the previous step and converts these to a Markdown format. The code for this was written in Python, is open-source and can be found in the *test-case-repository-word2md* repository [5]. In a nutshell, the script parses the Word files and translates any text into a Markdown format [8]. Any figures embedded in the Word file are extracted and saved in the same folder as the document itself. Once the conversion is done, a new release is created, and the next repository is triggered. The next repository in the workflow is the *test-case-repository* [3]. When this is triggered, it downloads the Markdown files and the figures from the *test-case-repository-word2md* repository and converts these into HTML files. For this step Hugo[4] is used. This tool simplifies the conversion of Markdown files into a

---

[1] https://github.com/.

[2] https://github.com/features/actions.

[3] https://pages.github.com/.

[4] https://gohugo.io/.

fully functional website using templates. After conversion, the website is hosted on GitHub Pages. You can find more details in the public repository linked below.

3.3 Publishing Test Case Descriptions

TCs provide useful information for users to implement laboratory tests. Therefore, it is important that previously created TCs can be easily shared between users. To maximize the long-term utility of the TCs developed within ERIGrid 2.0, an online repository was established. Also, the original format using Word files makes it difficult to find specific TCs related to a certain topic since they are not easily searchable. The newly established online repository allows users to publish their TCs by uploading Word files to a GitHub repository, where they are converted into web pages and hosted for broader access. This repository ensures that the TCs will remain accessible and beneficial to users even after the conclusion of the project. Furthermore, since the repository will stay active after the project ends, it will serve as a common repository for future developed TCs. The current version of the ERIGrid 2.0 Test Case Repository can be found here [2].

4 Future Work

Besides all the above outlined improvements, potential future work on the HTD method could focus on harmonizing it with other testing approaches to ensure seamless interoperability across diverse systems. Especially for interoperability testing, such a harmonization should take place between existing approaches and concepts, as highlighted in the int:net harmonisation report [14, 15]. Additionally, expanding the provision of TCs and sharing it with the online repository to cover a broader range of scenarios can enhance the robustness of the HTD framework as well.

Another potential activity is on the improvement of related software tools and the creation of more user-friendly interfaces to make the methodology more accessible to a wider audience, facilitating its adoption and implementation. Furthermore, developing approaches for the automated generation of HTD descriptions based on specific use cases and system configurations has the potential to streamline the testing process, reducing manual effort and increasing efficiency. These advancements will collectively contribute to a more comprehensive and effective testing framework for CPES. Finally, the international standardization of the HTD method could facilitate a further adoption of this approach (cf. Chap. 10).

References

1. Holistic test description templates, erigrid (2019) https://github.com/erigrid/holistic-test-description. https://github.com/ERIGrid/Holistic-Test-Description
2. Erigrid 2.0 test case repository (2024). https://erigrid2.github.io/test-case-repository/. Accessed 10 Feb 2024
3. Erigrid2/test-case-repository (2024). https://github.com/ERIGrid2/test-case-repository. Accessed 10 Feb 2024
4. Erigrid2/test-case-repository-word-input (2024). https://github.com/ERIGrid2/test-case-repository-word-input. Accessed 10 Feb 2024
5. Erigrid2/test-case-repository-word2md (2024). https://github.com/ERIGrid2/test-case-repository-word2md. Accessed 10 Feb 2024
6. D'Arco S, De Paola A, Widl E, Rajkumar VS, Kamsamrong J, Raussi P, Arnold G, Thomas D, Marinopoulos A, Wild CW, Heussen K, Rikos E, Hoang TT, Cortés-Borray AF (2022) D-jra1.1 benchmark scenarios. Technical report. https://doi.org/10.5281/zenodo.4032691
7. Widl E (2020) An overview of the precise approach used for the definition of simulation and optimization workflows. Technical report, AIT. https://ec.europa.eu/research/participants/documents/downloadPublic?documentIds=080166e5cd070ea9&appId=PPGMS
8. Gruber J (2004) Markdown 1.0.1. https://daringfireball.net/projects/markdown/. Accessed 10 Feb 2024
9. Heussen K, Steinbrink C, Abdulhadi IF, Nguyen VH, Degefa MZ, Merino J, Jensen TV, Guo H, Gehrke O, Bondy DEM, Babazadeh D, Pröstl Andrén F, Strasser TI (2019) Erigrid holistic test description for validating cyber-physical energy systems. Energies 12(14):1–31. https://doi.org/10.3390/en12142722
10. Petra R, Pröstl Andrén F, Heussen K, Edmund W (2024) D-n4.3 extended test description method. https://doi.org/10.5281/zenodo.14234745
11. Raussi P, Opas M, Strasser TI, Widl E, Kazmi J, Hoang TT, Tran QT, Rikos E, Khavari A, Heussen K, Gehrke O, Paspatis A, Gilbert I, Castro F, Kamsamrong J, Pellegrino L, Zerihun TA, Rajkumar VS (2021) Common reference test case profiles. Technical report, VTT Oy (VTT). https://doi.org/10.5281/zenodo.5522373. Version Number: 1.5
12. Schwarz JS, Schulte E, Heussen K, Nikoletatos J, Ramos L, Feng Z (2024) D-JRA1.2 methods for holistic test reproducibility. https://doi.org/10.5281/zenodo.8081442
13. Strasser T, de Jong E, Sosnina M (2020) European guide to power system testing: the ERIGrid holistic approach for evaluating complex smart grid configurations. Springer, Berlin. https://doi.org/10.1007/978-3-030-42274-5
14. Strasser T, Widl E, Kuchenbuch R, Lázaro-Elorriaga L, Tellado Laraudogoitia B, Ginocchi M, Penthong T, Ponci F, Gyrard A, Kung A, Mac Gregor C, Garcia Montero C, Relano Algaba E (2024) Towards interoperability testing of smart energy systems – an overview and discussion of possibilities. In: IET conference proceedings, vol 2024. https://doi.org/10.1049/icp.2024.4670
15. Widl, E., Strasser, T., Mac Gregor, C.A., Kuchenbuch, R., Lázaro Elorriaga, L., Tellado Laraudogoitia, B., Relano Algaba, E., Garcia Montero, C., Ginocchi, M., Penthong, T., Chy, T., Kung, A.: D3.1 testing concepts and procedures harmonisation report (2024). https://doi.org/10.5281/zenodo.15087499

Open Access This chapter is licensed under the terms of the Creative Commons Attribution 4.0 International License (http://creativecommons.org/licenses/by/4.0/), which permits use, sharing, adaptation, distribution and reproduction in any medium or format, as long as you give appropriate credit to the original author(s) and the source, provide a link to the Creative Commons license and indicate if changes were made.

The images or other third party material in this chapter are included in the chapter's Creative Commons license, unless indicated otherwise in a credit line to the material. If material is not included in the chapter's Creative Commons license and your intended use is not permitted by statutory regulation or exceeds the permitted use, you will need to obtain permission directly from the copyright holder.

Enhanced Validation Methods and Benchmark

J. Kamsamrong, E. Widl, J. S. Schwarz, K. Heussen, P. Raussi, G. Arnold, A. De Paola, Z. Feng, O. Werth, M. C. Pham, and Q. T. Tran

Abstract This chapter introduces enhanced validation methods and benchmarks for cyber-physical energy systems, focusing on reproducibility and uncertainty man-

---

J. Kamsamrong (✉) · J. S. Schwarz · O. Werth
OFFIS – Institute for Information Technology, Oldenburg, Germany
e-mail: jirapa.kamsamrong@offis.de

J. S. Schwarz
e-mail: jan.soeren.schwarz@offis.de

O. Werth
e-mail: oliver.werth@offis.de

E. Widl
AIT Austrian Institute of Technology, Vienna, Austria
e-mail: edmund.widl@ait.ac.at

K. Heussen
Technical University of Denmark, Kongens Lyngby, Denmark
e-mail: kheu@dtu.dk

P. Raussi
VTT Technical Research Centre of Finland, Espoo, Finland
e-mail: petra.raussi@vtt.fi

G. Arnold
Fraunhofer Institute for Energy Economics and Energy System Technology IEE, Kassel, Germany
e-mail: gunter.arnold@iee.fraunhofer.de

A. De Paola
Joint Research Centre of the European Commission, Ispra, Italy
e-mail: antonio.de-paola@ec.europa.eu

Z. Feng
University of Strathclyde, Glasgow, UK
e-mail: zhiwang.feng@strath.ac.uk

M. C. Pham · Q. T. Tran
French Alternative Energies and Atomic Energy Commission, Paris, France
e-mail: minh-cong.pham@cea.fr

Q. T. Tran
e-mail: quoctuan.tran@cea.fr

© The Author(s) 2025
T. I. Strasser et al. (eds.), *European Guide to Smart Energy System Testing*, SpringerBriefs in Energy, https://doi.org/10.1007/978-3-031-99451-7_3

agement. It presents benchmarks for electrical networks, multi-energy networks, and ICT-enhanced power systems, facilitating comprehensive testing and validation. Additionally, it highlights the contributions of the ERIGrid 2.0 project in developing these benchmarks and methods to support open science and improve system-level validation.

1 Introduction

CPES are complex, involving multiple domains and dependencies between different systems. This complexity poses significant challenges in managing the variability for testing and validation. Benchmarks serve as valuable tools for students, researchers, and industry practitioners by facilitating the testing of CPES. Existing reference models and lab setups work for specific small-scale experiments but face challenges for testing system-level hypotheses, for upscaling to real systems, or for extending to analyses to other domains (e.g., from the electricity grid to heat or communications). Another common challenge in validation is the recognition and declaration of uncertainties, which often hinders result reproducibility. A systematic approach for analyzing and accounting of uncertainty factors would mediate this challenge.

This chapter introduces the benchmark models developed in ERIGrid 2.0 and methods toward reproducible systems validation for CPES. The developed benchmarks and methods facilitate the methodical design and documentation of experiments, supporting open science and the further development of open-source paradigms [7].

2 Cyber-Physical Energy Systems Benchmarks

2.1 Concept and Approach

Modern energy systems have become increasingly complex, integrating multiple domains and introducing interdependencies. The shift from a centralized to a decentralized structure has underscored the need for ICT integration to enable real-time monitoring, enhanced control, and efficient data processing. This is crucial not only within a single sector but also across interconnected systems (e.g., Power-to-Heat) ensuring seamless energy flow management across domains. This advancement in energy systems creates the need for improving existing benchmarks and developing new benchmarks representing CPES for testing and validation.

The first question that may arise is: What are the differences between a model and a benchmark? Several power system models are published and available, such as the SIMBENCH, CIGRE, and IEEE families of reference models. *A model* refers to a mathematical or computational representation of a system (sometimes already implemented as source code or in binary form) for simulation and analysis. *A benchmark* in this context is a reference or standardized model used for performance evaluation and comparison on a certain application context, while *Benchmarking* refers to the process of comparing solutions on the benchmark or reference model. Thus, benchmarks can be valuable tools for knowledge consolidation through comparative analysis, as well as for testing and validation in CPES.

The next question then emerges: Which benchmark is suitable for testing in CPES? Existing models address primarily the power system, which may be insufficient when addressing the interdependencies for the CPES. In the context of CPES, functional scenarios serve as an umbrella that describes system functionalities, motivations, use cases, TCs, experimental setups, and their relevance. These scenarios provide a foundation for defining high-level requirements necessary for testing and simulation. High-level use cases can help to define motivation and requirements, while scenarios help outline possible system behaviour and interactions. Several test cases can be developed following the six Functional Scenarios introduced in chapter "Holistic Smart Energy System Validation", representing key technological areas for CPES. Although it is not feasible to have benchmarks that directly present every TC, relevant benchmarks can potentially be adapted to relevant TCs. Considering the development of CPES, which integrates ICT, power electronics, renewable energy sources, and sector coupling, three major benchmark categories are defined:

1. *Electrical Network*, representing traditional and modern power grids, including distributed energy integration [1].
2. *Multi-Energy Networks*, covering sector coupling, such as power-to-gas, power-to-heat, and integrated energy systems [2].
3. *ICT-Enhanced Power Systems*, addressing the role of communication networks, cybersecurity, and data exchange in modern power grids [3].

These benchmarks provide a structured foundation for testing and validation across different domains, ensuring comprehensive assessment and adaptability to emerging CPES challenges.

2.2 Electrical Network Benchmark

The Electrical Network Benchmark has been designed to provide a versatile testing platform for low-voltage grids in modern power system studies with high penetration of power electronics and renewable energy resources. Several iterations of the model have been considered to maximize the model's flexibility and strike an optimal trade-off between complexity and simulation speed. The benchmark is derived from several functional scenarios to address, especially electrical issues associated with converters operation, integration of renewables, and control schemes. Differently from the other proposed model, this benchmark focuses on implementations on a single software platform. The functional scenarios relevant for this benchmark are FS1 related to DER ancillary services, FS2 related to microgrids and energy communities, FS4 related

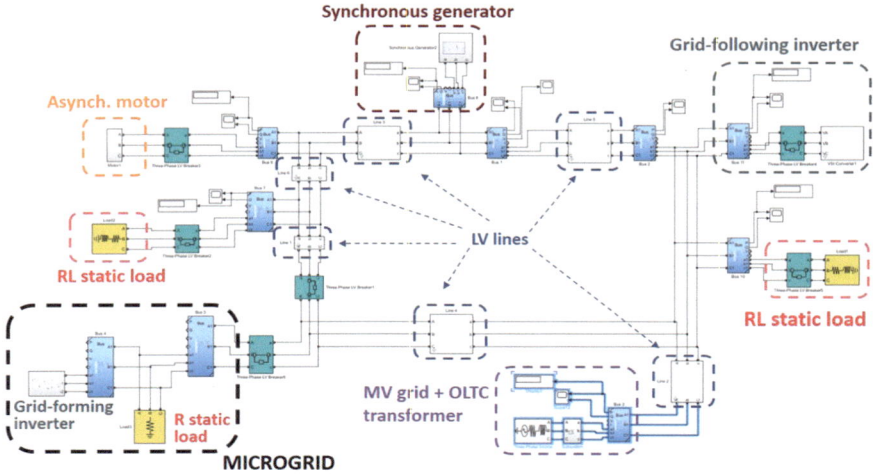

Fig. 1 Overview of the electrical network benchmark

to frequency and voltage stability, and FS5 related to aggregation and flexibility management (see also chapter "Holistic Smart Energy System Validation").

The benchmark includes a sufficiently complex network with an adequate number of buses and lines, circuit breakers for analyzing disconnections and topology changes, and inverters as controllable voltage sources with control schemes. The latter can also be used to represent a microgrid and assess the islanding capabilities of realistic systems. Additionally, at least two distinct DER types, such as PV systems and batteries, must be included to evaluate their interactions with synchronous generation. A distribution MV/LV transformer with an on-load tap-changer should also be incorporated to analyze its performance under varying system conditions. This has resulted in the inclusion of the following components in the benchmark, as shown in Fig. 1.

Examples of how this benchmark can be used include, but are not limited to:

- Testing the grid-forming capabilities of microgrid inverters.
- Voltage control with on-load tap changer controller.

The interested reader can find a detailed description of the model design, documentation and testing in [6].

2.3 Multi-energy Network Benchmark

The Multi-Energy Networks benchmark is a reference setup for sector coupling, where a power-to-heat facility connects a low-voltage electrical grid to a local heating network. An overview of the system configuration is presented in Fig. 2.

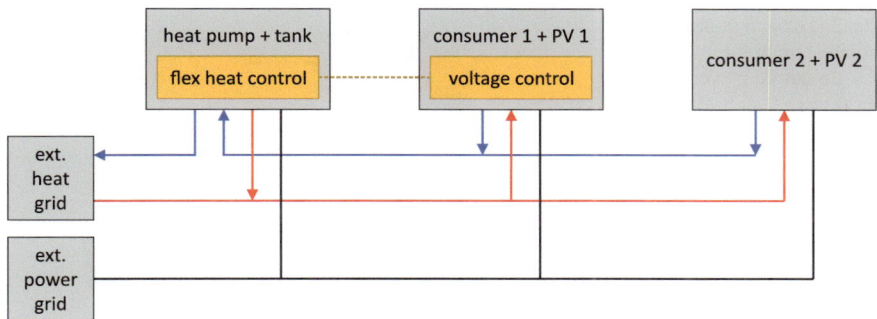

Fig. 2 Overview of the multi-energy networks benchmark [4]

Excess PV generation is used to improve grid stability and support thermal supply. From the perspective of the functional scenarios, this benchmark clusters together all sector coupling-related topics and specifically considers co-simulation for the assessment of the system. The relevant functional scenarios for this benchmark include FS2 related to microgrids and energy communities and FS3 related to sector coupling (see also chapter "Holistic Smart Energy System Validation").

The motivation of this benchmark is to promote research and development in thermal-electrical sector coupling by offering a not-too-complex yet practical reference model. It encourages co-simulation for the analysis of multi-domain systems (power, heat, control) and provides two implementation approaches. Unlike traditional simulation benchmarks, it does not provide numerical validation but serves as a conceptual guide.

Examples of how this benchmark can be used include, but are not limited to:

- Characterizing the availability of power-to-heat services and assessing their impact on the networks.
- Verifying the enhanced self-consumption of renewable energy sources (RES) in a coupled heat and power network through the use of power-to-heat technology.
- Scaling analysis of the heat storage considering sector-coupling effects [9].

2.4 ICT-Enhanced Power Systems Benchmark

The traditional power system has been transformed into a CPES, introducing interdependencies between electric power and communication systems. The performance of one system can directly impact the other. For example, a robust communication network can enhance data exchange in the power grid, while risks can also propagate between them, such as cascading communication failures leading to wide-area blackouts. As fast, reliable bidirectional communication is mandatory for real-time monitoring and control, this benchmark serves as a reference for a CPES setup consisting of the power system with ICT-based communication. It represents the impact of ICT

Fig. 3 General system architecture of the ICT-Enhanced power system benchmark

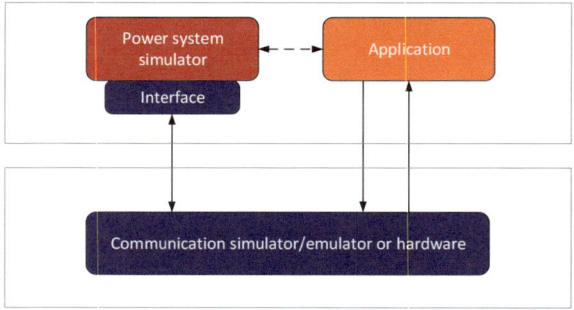

on power systems, focusing on communication networks that could hinder the control algorithms of power system components. The benchmark is aligned with functional scenario FS6 (see chapter "Holistic Smart Energy System Validation") on digitalisation and also encompasses other functional scenarios where ICT infrastructure or communication networks are essential for electric power system operation, enabling the coordination of distributed assets, substation automation, interoperability, and cybersecurity.

A general simulation setup for this benchmark is shown in Fig. 3, which consists of an electric power system simulator, communication simulator, tool/platform for hosting the application, co-simulation tool for information exchange and data connectivity between power system simulator, communication simulator and the application platform.

The use case for this benchmark involves but is not limited to:

- Real-time co-simulation models for power system and ICT interactions such as coordination of distributed assets, voltage control and protection.
- Assessing the reliability and latency of communication networks in CPES.
- Evaluation of communication performance for new CPES components and associated communication technologies.

3 Toward Reproducible Systems Validation for CPES

Two fundamental challenges to reproducibility are complexity and uncertainty. Complexity in CPES increases in several ways. For one, domain-extension introduces new cross-domain interaction phenomena that are not possible to reproduce and represent without including all relevant dynamics across several domains of expertise. Further, complexity results from a trend of scaling up by larger numbers of devices and subsystems (scaling out), which challenges the controllability of systems but also challenges the validation methods and tooling. Important practices for achieving reproducibility have been introduced above: the cross-domain benchmarks, the documentation of testing procedures using the HTD (see chapter "Holistic Smart Energy System Validation" and [11]), PreCISE for documentation of model details [5, 8], along with making models available as open-source implementations. Due to the complexity of CPES, another crucial topic for improving reproducibility is the handling of uncertainty in testing and validation, which needs to be addressed by specific methods.

In the following, the repertoire of methods and outline of a framework to tackle variability, sensitivities, scalability, and domain extension, which is described in more depth in [9, 15], is extended.

3.1 Uncertainty Concepts in Testing

Different types of uncertainty occur in the context of experiments. Those can result from a lack of knowledge (epistemic uncertainty) or from intrinsic fluctuations (aleatory uncertainty) and appear at different stages, for example, associated with input data, models, measurements, or the laboratory equipment. An approach to tackle such uncertainties can be described by three steps [15]. First, in the *uncertainty analysis* the uncertainties have to be identified and classified. Part of this first step, one can asses relative importance of uncertainties by quantifying the sensitivity of output measures to the uncertain factors. Second, in the *uncertainty representation*, a suitable approach for mathematically describing the uncertainty has to be found. Third, in the *Uncertainty Quantification (UQ)* the effect of uncertainties on the output of a system is calculated based on different approaches.

For smaller systems, an *analytical approach* might be suitable for UQ. A mathematical description of the system is needed, which can be used to propagate uncertainty through the system. For complex CPES, this is usually not possible, and *sampling approaches* might be more suitable. Those are repeating an experiment multiple times with changing input values and can be used for UQ and *Sensitivity Analysis (SA)*. SA here refers to a versatile set of methods, including local and global approaches, which can also be used for scaling analysis, as well as for creating an importance ranking of factors to identify the most impactful ones for more detailed investigation. More accurate UQ can be performed by explicitly propagating the uncertainty through a system. But this approach requires more knowledge about the uncertainties as described in [16] for stationary simulation models in the context of the co-simulation framework mosaik.

An extension of the HTD approach (see chapter "Holistic Smart Energy System Validation" and [11]) was developed to more directly integrate uncertainty analysis (see Sect. 2.3 and [10, 12]). This extension contains the Uncertainty Structure Analysis Tool (USAT), which is an Excel template with 4 structured sheets supporting the different steps of identification and prioritization of uncertainties within an experiment. The USAT integrates features to support the prioritization of factors for both uncertainty and scaling analysis purposes. Additionally, the template was extended in the following aspects:

- *Test Case (TC)*: New field for detailed Purpose of Investigation (PoI) and factor analysis alongside existing variability and quality attributes.
- *Qualification Strategy (QS)*: Includes uncertainty identification and management.
- *Test Specification (TS)*: Merges input/output parameters with sources of uncertainty, linking to the detailed USAT.
- *Experiment Realisation*: Links to USAT to assess uncertainty trade-offs.
- *Experimental Specification (ES)*: New fields for experimental setup uncertainties, precision of equipment, measurement uncertainty, and uncertainty management, and is also linking to USAT.

Based on the generic domain-independent method of description of System Configurations (SCs) introduced with the original HTD, the USAT is also well-suited to analyse multi-domain configurations. The scaling-up analysis is demonstrated in [9, 10].

3.2 Tool Chain

For the facilitation of the approaches described in the previous section, tools were developed in ERIGrid 2.0. Especially, the extension of the HTD for handling uncertainty aspects as described in Sect. 2.3 and a Design of Experiments (DoE) toolbox [14] can play together to implement a holistic tool chain for handling uncertainties in experiments. The DoE toolbox aims to support users in UQ and SA of experiments by providing a structure for the object-oriented description of the parameterization and variations and performing sample generation based on this. It, therefore, provides a complete parameterization for the recommended experiment runs and also supports in analysis of the results and the effects of the variations on the inputs. The usage of the USAT is explained on the multi-energy networks benchmark (see Sect. 2.3). In Fig. 4, the SC sheet is shown, which contains a list of parameters from the multi-energy benchmark.

For each parameter, information regarding its uncertainty can be collected, and the impact of the different parameters can be ranked with the help of the DoE toolbox, as described in [13]. Exemplarily shown are the distribution lines and the load, which were chosen as potential factors for the factor analysis (i.e., yellow column "(1) Potential factor?"), mapped to concrete experimental parameters in the implementation of the multi-energy networks benchmark (i.e., red column "(4) Mapping to experimental parameter and range"), and the results of a screening analysis was inserted (i.e., red column "(3) Factor ranking"). With those steps, the DoE toolbox can be used together with the HTD process. The USAT and DoE toolbox have the potential to be directly integrated in the future so that the parameterization for the DoE toolbox can be directly exported from the USAT. Such automation would also support the potential systematic investigation of quantitatively more complex configurations that arise, especially for scaling out test scenarios.

Component					Range		Type of uncertainty	Randomness or Lack of Knowledge/Data? (automatic suggestion)	Uncertainty Representation Type	Short explanation	1) Potential factor?	PoI Factor Analysis			
Component Name (automatic)	SC Subsystem ID	Parameter Name	Unit	Default Value	min	max						PoI #	PoI Target metric (automatic)	2) Factor in screening?	3) Factor ranking	4) Mapping to experimental Parameter and Range
Distribution line	1.2	Network parameters					seasonal variability and spatial uncertainty	epistemic and aleatory	Distribution	The variability of the resistance and reactance matrices due to ...	x	2	Voltage at consumer connection points (Vbus,i) ; Self-consumption ratio (SCratio)	x	15; 4	el_network.line_0_length; el_network.line_1_length
bus	1.3	Fault					uncertain initial state	epistemic and aleatory		Fault at the bus can...			N/A			
load	1.4	Time-dependent load demand variation					seasonal and random variability	epistemic and aleatory		Variation of load, type of load (const power, zip, or dynamic)	x	2	Voltage at consumer connection points (Vbus,i) ; Self-consumption ratio (SCratio)	x	20	consumer_load.scale

Fig. 4 Uncertainty structure analysis tool (USAT) SC parameter

4 Way Forward

Researchers use laboratory experiments and computer simulations to test hypotheses, but such results often lack reproducibility. The ERIGrid 2.0 project addressed these challenges by providing guidelines on handling uncertainties and scalability to enhance testing and validation for CPES. The developed methods and toolchain have been demonstrated and offer opportunities for further integration and automation. To support the analysis of scaling-out cases, for example, configurations will need to be created using algorithmic approaches. The HTD, USAT, and generic system configuration approach offer a useful semantic basis for developing interoperable and scalable toolchains for test specification, experimentation and analysis.

The three benchmarks and validation methods were developed in ERIGrid 2.0 following the FAIR principles ensuring findability, accessibility, interoperability, and reusability [17]. However, there remains room for improvement, particularly in enhancing interoperability. Future work should focus on advancing semantic interoperability and benchmark compatibility with multiple simulation platforms towards seamless integration across different research environments. These principles are not only supporting standardization efforts, but they are also fostering long-term sustainability and collaboration within the energy research community. Ongoing initiatives towards FAIR energy systems research will provide services and guidelines for the community that can serve as an orientation for benchmark compatibility.

References

1. Erigrid2/benchmark1 (2025). https://github.com/ERIGrid2/benchmark-model-electrical-network. Accessed 20 Mar 2025
2. Erigrid2/benchmark2 (2025). https://github.com/ERIGrid2/benchmark-model-multi-energy-networks. Accessed 20 Mar 2025
3. Erigrid2/benchmark3 (2025). https://github.com/ERIGrid2/benchmark-model-electrical-ict. Accessed 20 Mar 2025
4. CES A, Widl E (2021) ERIGrid2/benchmark-model-multi-energy-networks: v1.0. https://doi.org/10.5281/zenodo.5735005
5. D'Arco S, De Paola A, Widl E, Rajkumar VS, Kamsamrong J, Raussi P, Arnold G, Thomas D, Marinopoulos A, Wild CW, Heussen K, Rikos E, Hoang TT, CortÃs-Borray AF (2022) D-jra1.1 benchmark scenarios. Technical report. https://doi.org/10.5281/zenodo.4032691
6. De Paola A, Thomas D, Kotsakis E, Marinopoulos A, Masera M, Paspatis A, Kontou A, Kotsampopoulos P, Hatziargyriou N (2022) Benchmark models for low-voltage networks: a novel open-source approach. In: 2022 open source modelling and simulation of energy systems (OSMSES), pp 1–6. https://doi.org/10.1109/OSMSES54027.2022.9769097
7. De Paola A, Thomas D, Paspatis A, Widl E, Marinopoulos A, Kotsakis E, Kontou A, Kotsampopoulos P, Hatziargyriou N (2023) Testing-oriented development and open-source documentation of interoperable benchmark models for energy systems. IEEE Open J Ind Electron Soc 4:42–51. https://doi.org/10.1109/OJIES.2023.3234698
8. Widl E (2020) An overview of the precise approach used for the definition of simulation and optimization workflows. Technical report, AIT. https://ec.europa.eu/research/participants/documents/downloadPublic?documentIds=080166e5cd070ea9&appId=PPGMS

9. Heussen K, Schwarz JS, Cortés A, Feng Z, Nikoletatos J, Pham MC, Tran QT, Adrinanesis P, Kamsamrong J (2024) D-JRA1.3 methods for test upscaling and domain extension. https://doi.org/10.5281/zenodo.8081457
10. Heussen K, Schwarz JS, Schulte E, Feng Z, Ramos Perez LE, Nikoletatos J, Pröstl Andrén F (2025) Uncertainty annotations for holistic test description of cyber-physical energy systems. In: PowerTech 2025
11. Heussen K, Steinbrink C, Abdulhadi IF, Nguyen VH, Degefa MZ, Merino J, Jensen TV, Guo H, Gehrke O, Bondy DEM, Babazadeh D, Pröstl Andrén F, Strasser TI (2019) Erigrid holistic test description for validating cyber-physical energy systems. Energies 12(14):1–31. https://doi.org/10.3390/en12142722
12. Raussi P, Pröstl Andrén F, Heussen K, Edmund W (2024) D-N4.3 extended test description method. https://doi.org/10.5281/zenodo.14221590
13. Schwarz JS, Pham MC, Tran QT, Heussen K (2023) Scaling analysis in a multi-energy system. In: 2023 Asia meeting on environment and electrical engineering (EEE-AM), pp 01–06. https://doi.org/10.1109/EEE-AM58328.2023.10395068
14. Schwarz JS, Ramos Perez LE, Pham MC, Heussen K, Tran QT (2024) A toolbox for design of experiments for energy systems in co-simulation and hardware tests. In: 2024 open source modelling and simulation of energy systems (OSMSES), pp 1–7. https://doi.org/10.1109/OSMSES62085.2024.10668967
15. Schwarz JS, Schulte E, Heussen K, Nikoletatos J, Ramos L, Feng Z (2024) D-JRA1.2 methods for holistic test reproducibility. https://doi.org/10.5281/zenodo.8081442
16. Steinbrink C (2017) A non-intrusive uncertainty quantification system for modular smart grid co-simulation. Carl von Ossietzky Universität Oldenburg, Ph.D
17. Wilkinson MD, Dumontier M, Aalbersberg IJ et al (2016) The FAIR guiding principles for scientific data management and stewardship. Sci. Data 3(1):160018. https://doi.org/10.1038/sdata.2016.18

Open Access This chapter is licensed under the terms of the Creative Commons Attribution 4.0 International License (http://creativecommons.org/licenses/by/4.0/), which permits use, sharing, adaptation, distribution and reproduction in any medium or format, as long as you give appropriate credit to the original author(s) and the source, provide a link to the Creative Commons license and indicate if changes were made.

The images or other third party material in this chapter are included in the chapter's Creative Commons license, unless indicated otherwise in a credit line to the material. If material is not included in the chapter's Creative Commons license and your intended use is not permitted by statutory regulation or exceeds the permitted use, you will need to obtain permission directly from the copyright holder.

Extended Co-Simulation Approaches

J. S. Schwarz, E. Widl, O. Gehrke, K. Heussen, R. Fabian, and J. Kamsamrong

Abstract Co-Simulation plays an important role for testing new technologies for cyber-physical energy systems. This chapter describes different approaches for improving co-simulation in this context. Those approaches aim to improve initialization in co-simulation, event-based co-simulation of communication networks, multi-domain simulation including heat, and real-time co-simulation.

1 Cyber-Physical Energy Systems Co-Simulation Challenges

In the development of new technologies for CPES usually different steps are taken. With decreasing flexibility and increasing costs, the steps are: (i) mathematical analysis, (ii) software simulation, (iii) HIL-based simulation, (iv) laboratory tests, and

---

J. S. Schwarz (✉) · J. Kamsamrong
OFFIS – Institute for Information Technology, Oldenburg, Germany
e-mail: jan.soeren.schwarz@offis.de

J. Kamsamrong
e-mail: jirapa.kamsamrong@offis.de

E. Widl
AIT Austrian Institute of Technology, Vienna, Austria
e-mail: edmund.widl@ait.ac.at

O. Gehrke · K. Heussen
Technical University of Denmark, Kongens Lyngby, Denmark
e-mail: olge@dtu.dk

K. Heussen
e-mail: kheu@dtu.dk

R. Fabian
KEMA Labs, Arnhem, The Netherlands
e-mail: renzo.fabian@kema.com

© The Author(s) 2025
T. I. Strasser et al. (eds.), *European Guide to Smart Energy System Testing*,
SpringerBriefs in Energy, https://doi.org/10.1007/978-3-031-99451-7_4

(v) field tests [22]. Step ii aims to gain first knowledge about the integration into multi-domain systems with relatively small costs and prepare for more detailed but also expensive HIL, laboratory, and field tests.

For different domains, simulation models and tools exist and are established in the community but usually miss aspects of other domains. A common approach for the flexible integration of multiple domains is co-simulation, which allows to couple existing simulation models and tools in integrated simulation scenarios and avoids the need for new implementations of additional aspects from external domains. The coupled simulation components are usually executed concurrently and have to be synchronized, e.g., by a co-simulation framework. Depending on the use case, co-simulation can be executed as fast as possible or in real-time. Co-simulation with hard real-time requirements (*hard* means in this context at a level of milliseconds or less) can also be called "online", while simulation without those requirements can be called "offline". In general, software co-simulation can play an important role in the preparation of the following steps by doing a first evaluation of the results before implementing more expensive experiments. Therefore, in this chapter, the focus is on software simulation, while HIL simulation, which can be the next step after software simulation, is addressed in chapters "Improved Hardware-in-the-Loop-Based Testing" and "Laboratory Infrastructure Integration and Automation" of this book.

Many co-simulation frameworks are available with a focus on the simulation of the electrical grid, as shown in [12]. For the implementation in ERIGrid 2.0, mosaik was used in many cases because it is available as open-source software and well known in the community [23]. Especially, new features in mosaik 3, like event-based scheduling with the same time loops, were important improvements and allowed the implementation of new use cases in mosaik [13]. Compared to monolithic simulation, co-simulation still has some challenges due to the higher complexity, which comes with the integration of diverse simulation models and domains. One aspect is the initialization of simulation scenarios, which gets more complex due to diverse domains and potentially diverse modelling paradigms [26]. Two important facets of initialization are the way simulation scenarios can be defined and cyclic dependencies between the components in a simulation scenario.

Improvements for various aspects of co-simulation are described. The new event-based features of mosaik can enhance the efficiency of coupling communication and electrical networks, which is an important use case for co-simulation. Therefore, the implementation of a communication simulator based on the Functional Mockup Interface (FMI) is presented. To simplify the creation of new multi-domain simulation scenarios, multiple multi-domain benchmarks were developed. The technical implementation of them is explained. While all approaches mentioned so far focus on offline simulation, the last improvement discusses enhancements for real-time co-simulation.

2 Co-Simulation Setup and Initialisation

The configuration, invocation, and generic initialisation of co-simulation scenarios can be a challenge, which is important to tackle to facilitate multi-domain and improved scalable scenarios. The most important task of a co-simulation framework is the scheduling and data exchange of the coupled simulation components. In [25], improvements for the handling of cyclic dependencies in the co-simulation framework mosaik were described. Cyclic dependencies can lead to deadlocks in a co-simulation if two simulation components are waiting for data from the other components. In the predecessor ERIGrid, new mechanisms were introduced into mosaik to allow better definition of cyclic dependencies avoiding deadlocks. But those developments were not yet enough to allow frictionless initialisation of models across domains because their equilibrium conditions are not easily controllable since not all internal states are configurable for given black-box models. An option to deal with this problem is to extend the simulation time and throw away the results of the beginning to start the investigation with a reasonable simulation state, but this costs an additional amount of time, which could be omitted with improved temporal behaviour. Another important aspect regarding the configuration of simulation scenarios is the way they can be defined. For example, a graphical user interface can highly improve the usability of a simulation, but for better scalability, it is more important to have an option to define configurations in an interoperable format that offers the possibility of multiple analysis perspectives and interoperability between tools.

In the first version of the co-simulation framework mosaik, the scenario definition was done with a domain-specific language [16]. With the release of mosaik 2.0, this functionality was discarded to achieve better scalability and user friendliness with a Python-based scenario definition. Since then, mosaik provides a scenario Application Programming Interface (API), which can be used for flexible scenario creation, defining the simulation components, their parametrization, the data flows between them, and their temporal behaviour. In the release 3.0 of mosaik, extended approaches for handling cyclic dependencies with a superdense time concept were introduced [13]. It allows to define connections between simulation components, which can communicate multiple times without progress in the simulation time as visualised in Fig. 1. This can be helpful to implement the controller, but also for the

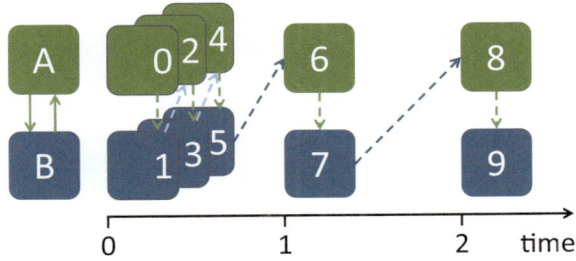

Fig. 1 Visualization of superdense time (i.e., same-time loops) in mosaik [14]

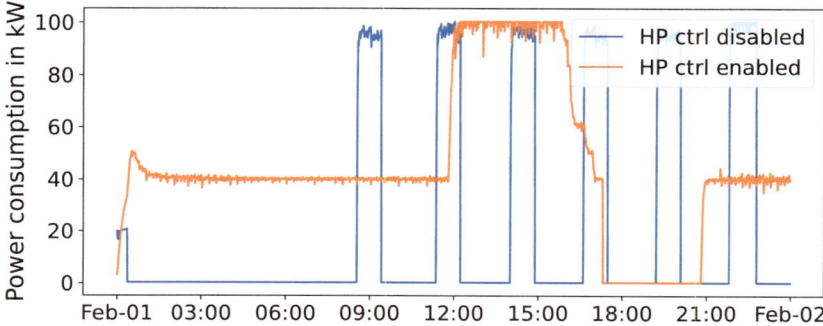

(a) Heat pump power consumption as simulated for one day; without same time loop.

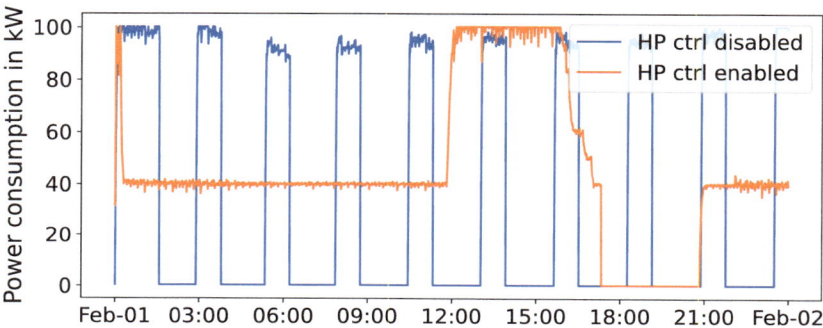

(b) Heat pump power consumption as simulated for one day; with same time loop.

Fig. 2 Comparison of simulation benchmark with and without same time loop [26]

initialisation of a scenario, because it can help to find an equilibrium between the states of multiple simulation models in the beginning of a simulation faster.

To give an example of the superdense time, an adapted version[1] of the multi-energy networks benchmark was developed [5, 6, 26]. As the scenario is quite complex and contains many cyclic dependencies, different sets of simulation components were tested as part of a same time loop, and the effects on the stability of the simulation and the results were investigated. The results show that the superdense time allowed to reach a reasonable simulation state significantly earlier and removing the beginning of the simulation results was not needed anymore. This can be shown based on the behaviour of a heat pump, which is part of the benchmark scenario. In the original version of the scenario, in the beginning, the heat pump didn't yet follow the typical temporal behaviour (see Fig. 2a). It needs almost nine hours of simulated time until the power pattern of the heat pump starts. In the adapted scenario, the

---

[1] https://github.com/ERIGrid2/JRA-2.1.3-STL.

initialisation of system states is achieved earlier, and the typical temporal behaviour occurs from the beginning (see Fig. 2b).

The superdense time implementation used for that example showed in its complexity with a high number of cyclic dependencies some problems in the stability, which were fixed in new mosaik updates. Additionally, the implementation still had some inconveniences because the same time loops were not transparently defined. Thus, since mosaik version 3.3, the *tiered durations* concept was introduced to further improve the same time loops in mosaik [15]. Before that, the simulation time was represented by an integer in mosaik and the same integer value for the simulation time was repeated in the same time loop. The new concept is based on the *tiered time* concept, which uses tuples of one or more integers to represent simulation time. Thus, any step can be subdivided by going one tier down. Additionally, the definition of consistent data flows based on tiered durations was simplified by using simulator groups.

The script/API-based formulation of mosaik offers convenient handles for automation, but the need to define and document simulation scenarios in a common data format remains. For domain-specific simulators (e.g. pandapower, MatPower) such file formats are well established and required to maintain and asses complex system configurations such as power grids. An extension for pandapipes was developed with this purpose.

3 Co-Simulation of Power Systems and ICT Networks

As traditional power systems evolve into CPES through the integration of ICT components to enhance communication and data exchange, testing and validation methods must also account for their interdependencies. Traditional monolithic simulation approaches are impractical for this use case and co-simulation can bridge the gap between different platforms or models. In Sect. 2.4, an ICT-Enhanced Power System benchmark was introduced, which connects DIgSILENT PowerFactory software representing the electrical grid and the communication emulator, Mininet, via an OPC Unified Architecture (OPC UA) interface. In contrast to this ad-hoc approach to cosimulation, the project also undertook a standards-based implementation of an integrated co-simulation of a power grid with communication, using the FMI[2] co-simulation standard and the mosaik platform.

The integration of power systems and ICT into an integrated simulation poses significant challenges due to their fundamentally different time bases: continuous-time behavior in power systems versus discrete, event-driven dynamics in communication networks. FMI serves as an open standard for model exchange (i.e., the exchange of entire models between simulators and a central solver) and co-simulation (i.e., the continuous exchange of state information between simulators), enabling seamless data exchange across different modeling tools. However, while earlier versions of the standard (e.g., FMI 2.0) allowed for variable timestep solvers for continuous-time

---

[2] https://fmi-standard.org/.

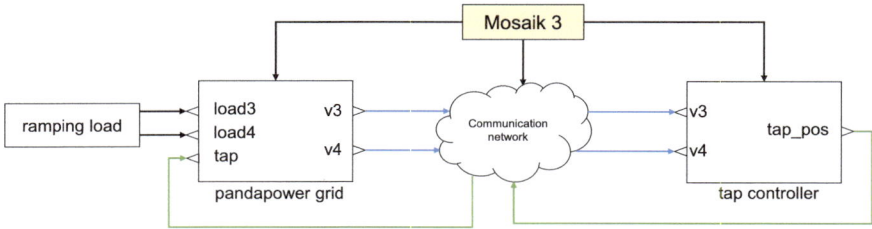

Fig. 3 Overview of co-simulation test setup based on mosaik 3 [6]

systems, there was no provision for event-driven systems in which individual simulators may not be able to complete a planned timestep due to an event occurring midstep. The recent release of FMI 3.0 addresses these shortcomings, providing a more comprehensive framework for co-simulation across diverse modelling paradigms.

The project intended the implementation to be a demonstration of these new capabilities, very shortly after the release of the FMI 3.0 standard, and when support of the new standard in simulation tools was still sparse to nonexistent. Consequently, FMI 3.0-compliant Functional Mockup Units (FMUs) had to be made from scratch. The project developed two variants of a simple communication network simulator, represented as an event-based queue/pipe with two possible behaviours: deterministic and probabilistic.

For the sake of simplicity, only the event-driven part of the simulation used FMI 3.0, as these were the only parts requiring the support for event-driven systems; the continuous-time simulators were coupled using FMI 2.0, which allowed the easy use of python-based simulators through the `pythonfmu` library for FMU generation. In this setup, the electrical grid was represented by `pandapower`. Additionally, a tap changer controller was implemented as a Python script.

Figure 3 shows the information exchange between the simulation components. `pandapower` performs a power flow calculation based on load input time series, determines the voltages at the load buses and transmits them via the communication network to the tap changer controller. The tap changer controller then calculates the optimal tap position and closes the loop by forwarding the result to `pandapower` (where the physical transformer is simulated) via the same communication network. The implementation of the scenario is described in more detail in [5], and the code is available at GitHub.[3]

A key outcome of this scenario is the FMI 3-compliant adapter for mosaik, which is described in more detail in [6]. It translates between the different time concepts of mosaik and FMI, manages the exchange of events between framework and coupled simulator, and allows the integration of FMI 3.0-based FMUs into mosaik 3.

---

[3] https://github.com/ERIGrid2/JRA2.1.1-PoC.

4 Heat Networks and Power System

Historically, investigations into multi-energy systems have predominantly concentrated on aligning supply and demand through the optimization of either scheduling or the operation of a large number of generation units [11]. Conversely, the examination of the effects on related infrastructure networks—such as the electrical grid, district heating system, and gas infrastructure—is a relatively recent subject [10, 21]. Simulation-based technical evaluations of multi-energy grids, primarily concentrating on operational and control, remain a challenge with the majority of available tools.

Recent years have seen advancements in the development of novel tools and methodologies for the integrated assessment of power and heat grids [8, 9, 20, 28]. The main objective of the approaches presented in the following was to showcase the potential of co-simulation for the technical assessment of coupled AC grids and thermal networks. This is not an exhaustive description of available tools but rather gives a hint at what innovative tools are capable of [27].

The primary component for modeling the thermal domain of multi-energy networks is the thermal network simulator. The network status is vital as it significantly influences heat supply and the functioning of coupling points with other sectors (e.g., power-to-heat). Consequently, the main requirement is to have a thermal network simulator that can be included in a co-simulation. The following gives 2 examples of very different types of simulators that can be used for this.

The Python package *pandapipes* offers a network calculation tool that may be used to analyze gas and district heating systems [9]. Pipe systems containing a fluid medium can be statically analyzed with the help of the pandapipes package. In district heating networks, this allows for the static or quasi-static calculation of pressure, temperature, and velocity distributions. Taking inspiration from electrical grid load flow calculations, the algorithm used to determine the network condition is called *pipe flow*. As a series of sequential snapshots of quasi-static thermo-hydraulic equilibrium states (steady-state operation), the evolution of the network state with time can be estimated.

The *DisHeatLib* is a tool for dynamically simulating district heating systems [8]. Thermo-hydraulic transients in pipe systems containing a fluid medium are the primary area of interest for the DisHeatLib library. This allows for the investigation of time-delayed propagation of fluid characteristics in the pipe system or flow reversals, two examples of dynamic effects in district heating networks. A non-linear pressure drop relation is used to represent the mass flow through all components in the so-called *plug flow* approach [24]. Components can be connected in any way desired, even in topologies with hydraulic loops or mesh networks, as the underlying thermo-hydraulic model's balance equations allow for this.

The subsequent advancements have been made to deliver a proof-of-concept implementation:

- The pandapipes package has been enhanced to conduct quasi-dynamic heat distribution simulations, which calculate dynamic temperature variations based on

pandapipes' steady-state hydraulic mass flow computations, taking into account the system's thermal inertia.
- The DisHeatLib library was enhanced to incorporate a heat pump model suitable for integration in a co-simulation with a power grid.
- The co-simulation configurations for the multi-energy network benchmark have been published and meticulously documented to function as reference implementations and to provide a foundation for additional training and tutorials.

Both the pandapipes package and the DisHeatLib library were applied to the same simulation use case (see Sect. 2.3). It comprises a low-voltage distribution network and a local branch of a heating network, connected through a power-to-heat facility. Figure 4 shows how the system being tested can be broken up into parts that can be represented by specialized simulators.

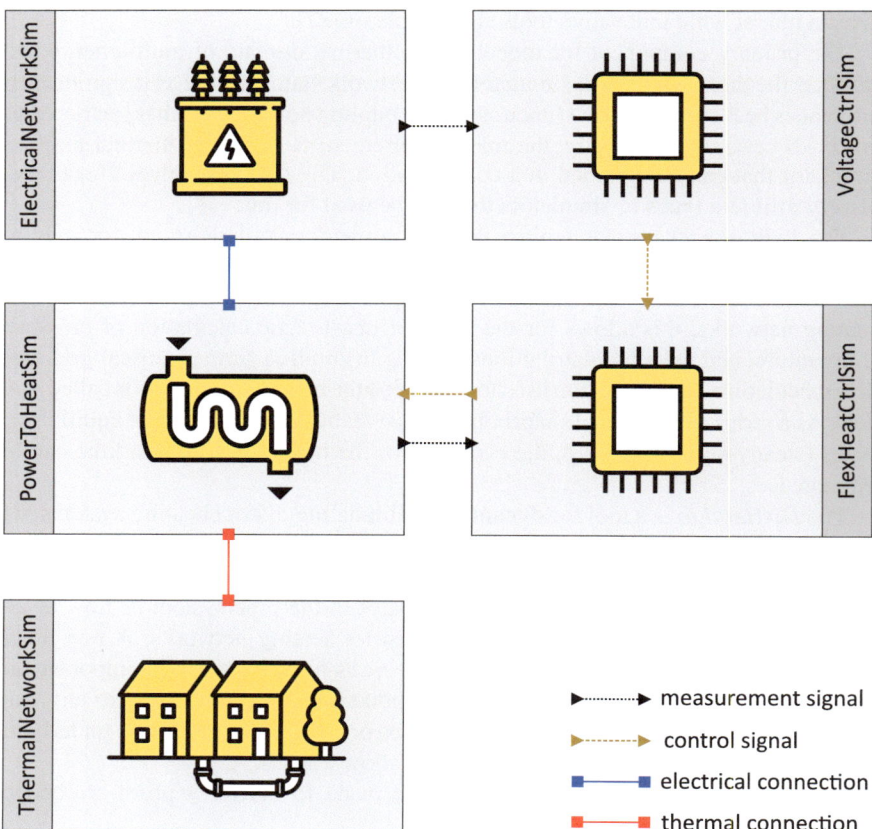

Fig. 4 Conceptual breakdown of the system under test into simulated subsystems for the multi-energy network benchmark [5]

However, the depicted partitioning is only conceptual; the different implementations combine or further split the shown blocks based on the features of the chosen simulation tools. For instance, for the parts of the thermal subsystem that pandapipes doesn't provide modeling (e.g., heat pumps and storage tanks), separate Python models are used. This creates a more fine-grained partitioning than shown in Fig. 4. On the other hand, when using the DisHeatLib, the entire thermal subsystem is modeled as one. Related to Fig. 4, this uses a simpler partitioning, where the thermal network and the power-to-heat facility are both represented by a single simulator.

The results of these co-simulation setups are discussed in Sect. 2. Even though both approaches still come with their limiting factors, they are expected to become more mature. Both approaches will provide viable, robust and openly available tools for the modeling of the thermal domain of multi-energy networks in the future.

5 Real-Time Co-Simulation Approach

Real-time co-simulation can be carried out in various ways, depending on criteria such as domain, sampling rate, and geographic location of the simulation platforms. The growing need for detailed analysis in large-scale networks has driven the development of diverse real-time co-simulation strategies. Early efforts in Electromagnetic Transients (EMT)-EMT co-simulation focused on establishing communication between RTDS and OPAL-RT systems using analog Inputs and Outputs (I/Os) for voltage and current exchange. More recently, digital protocols like Aurora have been adopted for lower latency and higher accuracy [2].

In Root Mean Square (RMS)-EMT co-simulation frameworks, the portion of the system requiring high accuracy is modeled using EMT methods, while the remaining parts are handled by RMS models. Attempting to model an entire system solely in EMT often faces constraints stemming from real-time hardware capacity, which can limit the size of the modeled circuit. To address these limitations, Thevenin equivalents are commonly used to reduce the model's complexity. However, if retaining detailed dynamics in certain segments is crucial, RMS-EMT co-simulation provides a viable alternative since phasor-based RMS simulations are less computationally intensive. Electromechanical phenomena can typically be simulated with time steps on the order of tens of milliseconds, whereas EMT simulations usually require microsecond-level steps. Several applications can be found in [3, 4, 7, 18, 19, 29].

Implementing such co-simulation approaches requires specific prerequisites, notably hardware interfaces capable of supporting reliable communication and synchronization between the real-time simulation platforms (e.g., analog I/O for voltage and current measurements or digital protocols such as Ethernet and Aurora). Robust coupling interfaces are also essential, similar to those employed in offline co-simulations, but with the added constraint that coupling algorithms must execute efficiently in real-time. Recent work has demonstrated that RMS-EMT co-simulation can achieve strong performance when latency and coupling requirements are managed effectively, enabling the simulation of numerous buses in the RMS domain while preserving a detailed EMT representation of critical segments. This approach has proven especially useful in CHIL applications, offering flexibility and efficiency for complex power system studies under realistic conditions.

In addition to purely electrical simulations, multi-domain co-simulation frameworks now include the communication network within the same environment [1, 17]. In such cases, the electrical system is simulated alongside the communication infrastructure, allowing researchers to capture the bidirectional impact of communication delays, data integrity, and network congestion on power system performance. This multi-domain perspective is valuable for advanced applications such as wide-area measurement systems or distribution automation, where communication reliability and latency directly affect control actions and overall system operation.

6 Final Thoughts

To enable more complex co-simulation of CPES, we have shown several approaches to improve the capabilities of co-simulation. First, an enhanced initialization, which uses same time loops, to significantly reduce the initialization phase of multi-domain scenario. Second, the new discrete-event features of FMI 3 and mosaik 3 were showcased in an example scenario integrating power grid and communication network. Third, the implementation of a multi-energy networks benchmark integrating heat networks and the power grid. Fourth, real-time co-simulation can improve performance by combining different detail levels in an integrated co-simulation.

The approaches provide new opportunities for using co-simulation, but further extensions might increase their impact in the future. For the same time loop-based initialization, the inclusion of a decision-making algorithm in the co-simulation configuration phase that intelligently designates the same time loop simulators to make optimally use of this functionality [26] would be interesting. Additionally, the simple FMI 3.0-based communication pipeline FMU might be extended in the future into a configurable network simulator and existing simulation tools such as ns-3 could be integrated.

Finally, real-time co-simulation encompassing EMT-EMT, RMS-EMT, and multi-domain approaches continues to advance as a pivotal technique for analyzing complex power systems. Ongoing developments in hardware interfaces, coupling algorithms, and communication protocols promise enhanced interoperability and flexibility in real-time simulation platforms, further enabling comprehensive and precise studies of future power grids.

References

1. Ali O, Mohammed OA (2024) Real-time co-simulation implementation for voltage and frequency regulation in standalone ac microgrid with communication network performance analysis across traffic variations. Energies 17(19). https://doi.org/10.3390/en17194872

2. Barbierato L, Pons E, Mazza A, Bompard EF, Rajkumar VS, Palensky P, Macii E, Bottaccioli L, Patti E (2021) Stability and accuracy analysis of a real-time co-simulation infrastructure. In: 2021 IEEE international conference on environment and electrical engineering and 2021 IEEE industrial and commercial power systems europe (EEEIC / I&CPS Europe), pp 1–6. https://doi.org/10.1109/EEEIC/ICPSEurope51590.2021.9584687
3. Fabián Espinoza R, Godoy P (2024) Opendss-based real-time RMS simulator: design and applications. In: 2024 open source modelling and simulation of energy systems (OSMSES), pp 1–7. https://doi.org/10.1109/OSMSES62085.2024.10668999
4. Fabián Espinoza R, Justino G, Otto RB, Ramos R (2021) Real-time rms-emt co-simulation and its application in Hil testing of protective relays. Electr Power Syst Res 197:107326. https://doi.org/10.1016/j.epsr.2021.107326
5. Gehrke O, Lauss G, Widl E, Rajkumar V, van der Meer A, Kontou A, Paspatis A, Vogel S, Tran The H, De Paola A, Syed M, Zhiwang F, Schwarz JS (2023) D-JRA2.1 coupling methods. https://doi.org/10.5281/zenodo.7958622
6. Gehrke O, Schwarz JS (2023) D-JRA2.2 software release. https://doi.org/10.5281/zenodo.7958642
7. Jalili-Marandi V, Dinavahi V, Strunz K, Martinez JA, Ramirez A (2009) Interfacing techniques for transient stability and electromagnetic transient programs IEEE task force on interfacing techniques for simulation tools. IEEE Trans Power Deliv 24(4):2385–2395. https://doi.org/10.1109/TPWRD.2008.2002889
8. Leitner B, Widl E et al (2019) A method for technical assessment of power-to-heat use cases to couple local district heating and electrical distribution grids. Energy 182:729–738. https://doi.org/10.1016/j.energy.2019.06.016
9. Lohmeier D, Cronbach D et al (2020) Pandapipes: an open-source piping grid calculation package for multi-energy grid simulations. Sustainability 12(23). https://doi.org/10.3390/su12239899
10. Lund H (2018) Renewable heating strategies and their consequences for storage and grid infrastructures comparing a smart grid to a smart energy systems approach. Energy 151:94–102. https://doi.org/10.1016/j.energy.2018.03.010
11. Mancarella P (2014) MES (multi-energy systems): an overview of concepts and evaluation models. Energy 65:1–17. https://doi.org/10.1016/j.energy.2013 10.041
12. Mihal P, Schvarcbacher M, Rossi B, Pitner T (2022) Smart grids co-simulations: Survey & research directions. Sustain Comput: Inform Syst 35. https://doi.org/10.1016/j.suscom.2022.100726
13. Ofenloch A, Schwarz JS, Tolk D, Brandt T, Eilers R, Ramirez R, Raub T, Lehnhoff S (2022) MOSAIK 3.0 : combining time stepped and discrete event simulation. In: 2022 open source modelling and simulation of energy systems (OSMSES). Aachen
14. Raub T, Schwarz JS, Ofenloch A, Brandt T (2021). Mosaik scheduling figures. https://doi.org/10.6084/m9.figshare.16988437.v1
15. Schulte E, Balduin S (2024) Tiered durations: scheduling at differing time resolutions. In: The 38th annual European simulation and modelling conference 2024. San Sebastian
16. Schütte S (2013) Simulation model composition for the large-scale analysis of smart grid control mechanisms. Ph.D. thesis, Carl von Ossietzky University Oldenburg
17. Shi K, Ren X, Da T, Sun H, Zou D, Cheng F, Liu W (2021) Real-time co-simulation platform for cyber-physical power system based on 5g communication technology. In: 2021 IEEE sustainable power and energy conference (iSPEC), pp 4052–4057. https://doi.org/10.1109/iSPEC53008.2021.9735847
18. Sidwall K, Forsyth P (2020) Advancements in real-time simulation for the validation of grid modernization technologies. Energies 13(16). https://doi.org/10.3390/en13164036
19. Song J, Hur K, Lee J, Lee H, Lee J, Jung S, Shin J, Kim H (2020) Hardware-in-the-loop simulation using real-time hybrid-simulator for dynamic performance test of power electronics equipment in large power system. Energies 13(15). https://doi.org/10.3390/en13153955
20. Song R, Hamacher T, Perić VS (2021) Impact of hydraulic faults on the electric system in an integrated multi-energy microgrid. In: Proceedings of the 9th workshop on modeling and simulation of cyber-physical energy systems. https://doi.org/10.1145/3470481.3472709

21. Sorknæs P, Lund H, Andersen AN (2015) Future power market and sustainable energy solutions—the treatment of uncertainties in the daily operation of combined heat and power plants. Appl Energy 144:129–138. https://doi.org/10.1016/j.apenergy.2015.02.041
22. Steinbrink C (2017) A non-intrusive uncertainty quantification system for modular smart grid co-simulation. Carl von Ossietzky Universität Oldenburg, Ph.D.
23. Steinbrink C, Blank-Babazadeh M, El-Ama A, Holly S, Lüers B, Nebel-Wenner M, Ramírez Acosta R, Raub T, Schwarz JS, Stark S, Nieße A, Lehnhoff S (2019) CPES testing with mosaik: co-simulation planning. Exec Anal Appl Sci 9(5):923. https://doi.org/10.3390/app9050923
24. van der Heijde B, Fuchs M et al (2017) Dynamic equation-based thermo-hydraulic pipe model for district heating and cooling systems. Energy Convers Manag 151:158–169. https://doi.org/10.1016/j.enconman.2017.08.072
25. van der Meer AA, Bhandia R, Palensky P, Cvetković M, Widl E, Nguyen VH, Tran QT, Heussen K (2020) Simulation-based assessment methods. In: Strasser TI, de Jong ECW, Sosnina M (eds) European guide to power system testing: the Erigrid holistic approach for evaluating complex smart grid configurations. Springer International Publishing, Cham, pp 35–50. https://doi.org/10.1007/978-3-030-42274-5_3
26. van der Meer AA, Schwarz JS, Heussen K (2022) Qualification of initialisation challenges in co-simulation setups for integrated energy systems. In: 10th workshop on modelling and simulation of cyber-physical energy systems (MSCPES), pp 1–6. https://doi.org/10.1109/MSCPES55116.2022.9770106
27. Widl E, Wild C, Heussen K, Rikos E, Hoang TT (2022) Comparison of two approaches for modeling the thermal domain of multi-energy networks. In: 2022 Open source modelling and simulation of energy systems (OSMSES), pp 1–6. https://doi.org/10.1109/OSMSES54027.2022.9769129
28. Wirtz M, Remmen P, Müller D (2021) EHDO: a free and open-source webtool for designing and optimizing multi-energy systems based on MILP 29(5):983–993. https://doi.org/10.1002/cae.22352
29. Zhang Y, Gole AM, Wu W, Zhang B, Sun H (2013) Development and analysis of applicability of a hybrid transient simulation platform combining TSA and EMT elements. IEEE Trans Power Syst 28(1):357–366. https://doi.org/10.1109/TPWRS.2012.2196450

Open Access This chapter is licensed under the terms of the Creative Commons Attribution 4.0 International License (http://creativecommons.org/licenses/by/4.0/), which permits use, sharing, adaptation, distribution and reproduction in any medium or format, as long as you give appropriate credit to the original author(s) and the source, provide a link to the Creative Commons license and indicate if changes were made.

The images or other third party material in this chapter are included in the chapter's Creative Commons license, unless indicated otherwise in a credit line to the material. If material is not included in the chapter's Creative Commons license and your intended use is not permitted by statutory regulation or exceeds the permitted use, you will need to obtain permission directly from the copyright holder.

Improved Hardware-in-the-Loop-Based Testing

Z. Feng, M. H. Syed, A. Paspatis, A. Kontou, G. Lauss, A. De Paola, P. Kotsampopoulos, N. Hatziargyriou, and G. Burt

Abstract Real-time simulation and hardware-in-the-loop techniques are crucial for testing advanced power energy systems. This chapter covers the state-of-the-art of hardware-in-the-loop techniques, highlighting its key research and application challenges associated with stability, accuracy, and sensitivity. Advanced hardware-in-the-loop stability enhancement schemes, time delay compensation-aided accuracy improvement, and a novel framework for hardware-in-the-loop sensitivity analysis are presented. Based on these approaches, case studies showcasing successful

---

Z. Feng (✉) · G. Burt
University of Strathclyde, Glasgow, UK
e-mail: zhiwang.feng@strath.ac.uk

G. Burt
e-mail: graeme.burt@strath.ac.uk

M. H. Syed
WSP Energy Advisory, Glasgow, UK
e-mail: mazher.syed@wsp.com

A. Paspatis
Manchester Metropolitan University, Manchester, UK
e-mail: a.paspatis@mmu.ac.uk

A. Kontou · P. Kotsampopoulos · N. Hatziargyriou
ICCS, National Technical University of Athens, Athens, Greece
e-mail: alkistiskont@mail.ntua.gr

P. Kotsampopoulos
e-mail: kotsa@power.ece.ntua.gr

N. Hatziargyriou
e-mail: nh@power.ece.ntua.gr

G. Lauss
AIT Austrian Institute of Technology, Vienna, Austria
e-mail: georg.lauss@ait.ac.at

A. De Paola
Joint Research Centre of the European Commission, Ispra, Italy
e-mail: antonio.de-paola@ec.europa.eu

© The Author(s) 2025
T. I. Strasser et al. (eds.), *European Guide to Smart Energy System Testing*, SpringerBriefs in Energy, https://doi.org/10.1007/978-3-031-99451-7_5

hardware-in-the-loop testing of advanced power techniques are presented at the end of this chapter.

1 Hardware-in-the-Loop-Based CPES Testing

1.1 State-of-the-Art and System Modelling

HIL simulation integrates physical power components with a Real-Time Simulator (RTS) into a closed-loop system configuration, effectively replicating the original System of Interest (SOI). Figure 1 illustrates the SOI alongside its corresponding power HIL (PHIL) simulation setup. The SOI is represented by a voltage divider topology consisting of two interconnected Thevenin equivalents S_1 and S_2. The Model of Interest (MOI) S_1 corresponds to the real-time network within the digital RTS platform, while S_2 represents the Hardware of Interest (HOI). S_1 consists of a voltage source V_S in series with an equivalent impedance Z_S, representing the model implemented on an RTS. S_2 comprises the hardware equivalent impedance Z_H. As shown in Fig. 1b, these two subsystems are coupled through a PHIL interface, enabling real-time interaction between the MOI and the HOI.

A PHIL interface – in literature often called Power Interface (PI) as well – consists of a Power Amplifier (PA), signal conversion cards, and sensors. The configuration of these components and the method by which power is transferred between MOI and HOI are determined by specific Interface Algorithms (IAs), as depicted in [15]. Among these IAs, the Ideal Transformer Model (ITM) interface is widely adopted due to its straightforward implementation and strong performance in terms of stability and accuracy. As shown in Fig. 2a, the ITM interface consists of a controllable voltage source on the HOI side and a controllable current source on the MOI side. The voltage source is implemented as a PA that regulates its output voltage based on the command signal measured at MOI coupling point and transmitted via a Digital-to-Analog Conversion (DAC) card. The current on the HOI side is measured by a current sensor and processed through a Low-Pass Filter (LPF). The filter output is subsequently injected into the emulated network via a controlled current source, thus closing the test loop.

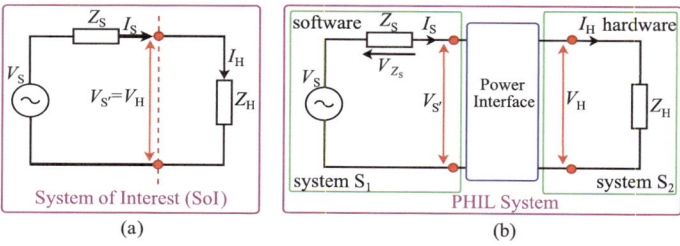

Fig. 1 Topology of **a** SOI and **b** the corresponding PHIL simulation system

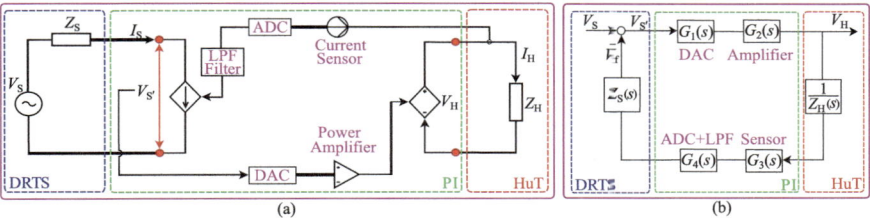

Fig. 2 **a** Equivalent model of PHIL setup and **b** its equivalent block diagram

From a system modelling standpoint, the PHIL system can be represented as a single-input-single-output system, whose equivalent block diagram, illustrating all key components and interface signals, is shown in Fig. 2b. The open-loop transfer function of this PHIL system is represented as

$$T_O(s) = \underbrace{G_1(s)G_2(s)G_3(s)G_4(s)}_{G_{PI}(s)} \frac{Z_S(s)}{Z_F(s)}, \quad (1)$$

where $G_{PI}(s)$ represents the power interface, which includes the transfer functions of DAC card $G_1(s)$, PA $G_2(s)$, current sensor $G_3(s)$, and Analog-to-Digital Conversion (ADC) card in series with an LPF $G_4(s)$. These components are modelled as

$$\begin{cases} G_1(s) = k_1 e^{-sT_{d1}}, \ G_2(s) = k_2 \dfrac{1}{\frac{s}{2\pi f_{c2}} + 1} e^{-sT_{d2}}, \\ G_3(s) = \ e^{-sT_{d3}}, \ G_4(s) = \ \dfrac{1}{\frac{s}{2\pi f_{c4}} + 1} e^{-sT_{d4}}, \end{cases} \quad (2)$$

where k_1 and k_2 are the DAC and PA scaling factors, respectively. T_{d1} and T_{d4} represent the time-step delay of RTS, T_{d2} and T_{d3} are the time delay of PA and sensor, respectively. Frequencies f_{c2} and f_{c4} are the corner frequency of PA and LPF, respectively [6]. The open-loop transfer function in Eq. 1 is further expressed as

$$T_O(s) = k_1 k_2 \frac{1}{\frac{s}{2\pi f_{c2}} + 1} \frac{1}{\frac{s}{2\pi f_{c4}} + 1} \frac{Z_S(s)}{Z_H(s)} e^{-sT_d} = G_{PI}^*(s) \frac{Z_S(s)}{Z_H(s)} e^{-sT_d}, \quad (3)$$

where $G_{PI}^*(s)$ represents the delay-free part of $G_{PI}(s)$ and T_d is the aggregated time delay of the interfacing components depicted in Eq. 2.

1.2 Research and Application Challenges

From both research and application perspectives, the key challenges hindering the robust and widespread adoption of the PHIL system can be summarized as follows.

System stability: The stability of the PHIL closed-loop system can be evaluated using the Nyquist criteria applied to the system characteristic equation (i.e., $1 + T_O(s) = 0$). Consequently, as highlighted in [6], Gain Margin (GM) and Phase Margin (PM) serve as key indicators of the PHIL stability. Apart from the intrinsic delay-free power interface $G_{PI}^*(s)$, the time delay T_d and the Impedance Ratio (IR) between the simulation and HOI side in the open-loop transfer function in Eq. 3 are the key determinants of stability, with GM and PM satisfying the Nyquist stability criteria. As illustrated in Fig. 3a, for a PHIL system with fixed impedance, the presence of time delay destabilises the system once it exceeds the threshold T_{dc}. Due to the inherent time delay and its associated stability challenges, the permissible IR range of the PHIL system is significantly constrained, limiting its broader application. Furthermore, as in Fig. 3b, for a PHIL system with fixed time delay, the increment in IR eventually breaches system stability. This stringent IR constraint to ensure a stable PHIL system poses a major challenge to the practical implementation of PHIL

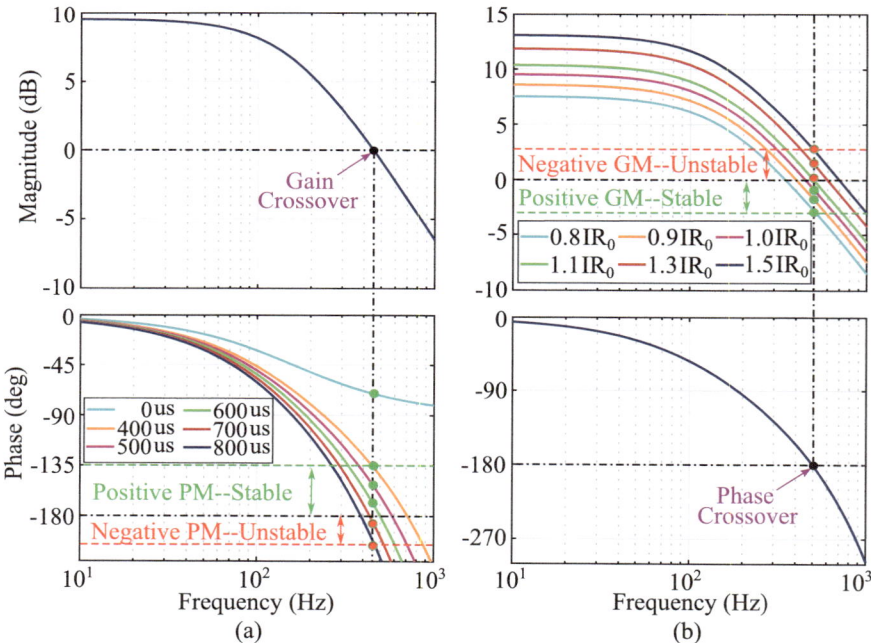

Fig. 3 a Bode plots of $T_O(s)$ of PHIL system with unity $G_{PI}^*(s)$ and (**a**) fixed IR and variable time delay, **b** fixed time delay ($T_d = 600us$) and variable IR [6]

setups, as real-time PHIL setups may experience significant variations in impedance on both the MOI and HOI sides during an experimental scenario run.

System accuracy: An ideal PHIL interface is delay-free and features transparent power amplification with unity gain and infinite bandwidth, allowing it to replicate the SoI depicted in Fig. 1a identically. However, in practice, the non-unity gain of the interfacing components and the phase lag introduced by the time delay lead to inconsistent magnitude and phase relationship between V_S and V_H in the PHIL system. This discrepancy disrupts accurate power signal synchronization between MOI and HOI and degrades the current response at both ends. Moreover, as demonstrated in [5], the phase shift (φ) arising from the time delay and PHIL interface alters the phase relationship between the voltage and current at both the MOI and HOI ends, leading to deteriorated power transfer between MOI and HOI sides. The relationship between the MOI power ($P_S + jQ_S$) and HOI power ($P_H + jQ_H$) is derived in [5] and is expressed as,

$$\begin{cases} P_H = |G_{PI}(s)|(P_S \cos\varphi - Q_S \sin\varphi), \\ Q_H = |G_{PI}(s)|(Q_S \cos\varphi + P_S \sin\varphi). \end{cases} \quad (4)$$

System sensitivity: Due to the incorporation of interfacing subsystems into the PHIL setup, their associated non-ideal characteristics not only significantly deteriorate PHIL stability and accuracy but also render the system more vulnerable to these imperfections, particularly to external disturbances that are inevitably introduced into the PHIL setup. These external disturbances originate from (i) offset in power signal measurement units, (ii) quantisation errors and noise in ADC and DAC signal conversion cards, (iii) switching harmonics arising from PA high-frequency pulsating modulation, and (iv) measurement noise in voltage and current sensors (typically high-frequency). From an application perspective, conducting a comprehensive sensitivity analysis, evaluating the impact of these disturbances on the PHIL setup, and identifying sensitivity issues caused by external disturbances are essential for achieving high-fidelity and robust PHIL simulations before final-stage deployment. In contrast to the well-established stability and accuracy assessment schemes presented in the literature, sensitivity analysis and assessment remain unexplored within the PHIL community, with no dedicated scheme developed before the research efforts [11] in ERIGrid 2.0 aimed at bridging this gap.

2 Hardware-in-the-Loop Simulation Improvements

2.1 Approaches for PHIL Stability Enhancement

To alleviate the destabilising impact of the key determinants identified above, numerous research efforts have focused on enhancing the PHIL system's stability. These approaches include refining the impedance ratio between the MOI and HOI ends

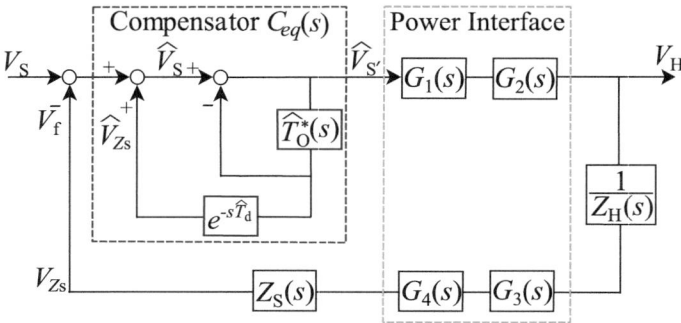

Fig. 4 PHIL diagram with detailed structure of SP compensator

[10] or increasing the system stability margins by augmenting the power interface with compensation schemes [12]. Building upon these advancements, the following novel PHIL stability enhancement schemes have been developed in ERIGrid 2.0.

Adaptive Smith Predictor-aided PHIL stability enhancement: To counteract the detrimental impact of the time delay on PHIL closed-loop stability, a Smith Predictor (SP) -based compensator is proposed in [4, 6]. As shown in Fig. 4, the SP compensator is integrated at the point of common coupling within the feed-forward path to manipulate the output signal from MOI before it is amplified by PA. The SP compensator functions by equivalently shifting the time delay out of the PHIL closed-loop, thereby enabling a delay-free PHIL interface with an enhanced stability margin. This improvement in stability margin extends IR, allowing for a higher IR threshold before breaching the system stability. Consequently, this approach introduces greater flexibility in impedance variations, enabling stable PHIL experiments across a broader range of operating conditions.

Furthermore, robustness analysis of SP compensator against the modelling error $\delta T_O^*(s)$ arising from system impedance variations is presented in [6]. As shown in Fig. 5a, the stability constraint imposed by the modeling error is quantified, serving as a key metric for assessing the robustness of the SP compensator. Accordingly, the critical impedance ratio IR_c, which defines the buffer for IR variability while maintaining stability in the SP-compensated PHIL system, is further quantified in Fig. 5b. This analysis reveals that while a passive SP compensator enhances stability by providing a buffer for impedance variations, it fails to maintain stability when the PHIL system witnesses a broader range of IR fluctuations.

To address the limitations of a passive SP compensator, as illustrated in Fig. 6, an adaptive SP compensator is further proposed in [6]. This approach leverages the Sliding Discrete Fourier Transform (SDFT) aided online impedance parameter identification technique, ensuring good agreement between the estimated and reference impedance transfer functions. This allows real-time adaptation of the SP compensator parameters, enabling stable and seamless operation across a broader range of impedance variations during experimental runs. This is essential for a robust PHIL experimental assessment of candidate HOI with naturally fluctuating impedance.

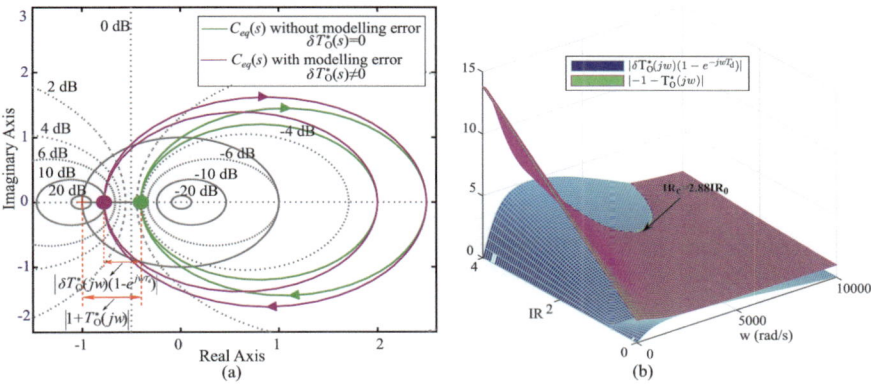

Fig. 5 **a** SP compensator robustness analysis against modelling error, and **b** critical IR quantification of passive SP based PHIL setup

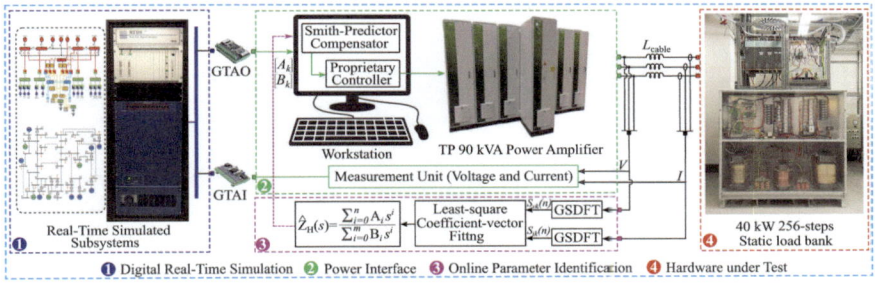

Fig. 6 PHIL experimental setup with adaptive SP compensator

Furthermore, this approach can be extended and applied to PHIL setups integrating parallel power electronics systems or physical power networks with dynamic impedance characteristics.

Virtual Shifting Impedance (VSI)-based PHIL stability enhancement: The Feedback Filtering (FBF) method was proposed in [12] to stabilise the PHIL setup at the expense of reduced system bandwidth and accuracy. Furthermore, the physical shifting impedance method was developed in [10] to shift a portion of MOI impedance to the HOI end. However, its applicability is limited by the dependence on the availability of hardware impedances of different values to conduct multiple test cases. To address these challenges, a novel Virtual Shifting Impedance (VSI) interface was developed as part of ERIGrid 2.0 [13]. Building upon existing IAs, this method enhances PHIL stability and accuracy by virtually shifting a portion of the MOI impedance to the HOI side. The detailed derivation and implementation of VSI are depicted in [13]. Compared to conventional approaches, as shown in Fig. 7, instead of relying on physical hardware impedance, this method manipulates the PA

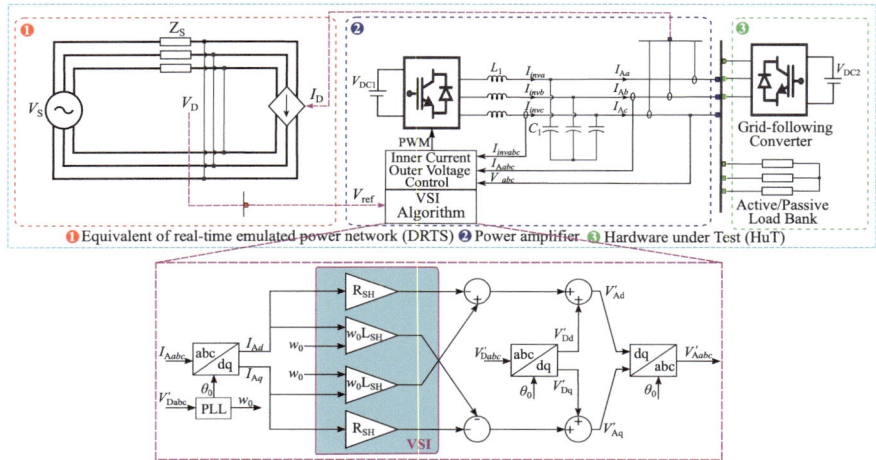

Fig. 7 PHIL equivalent block diagram implemented with VSI algorithm

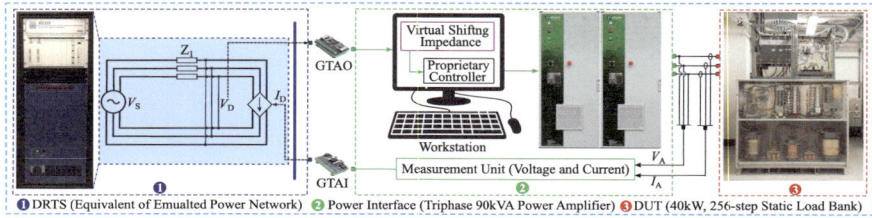

Fig. 8 PHIL setup for experimental validation of VSI algorithm

controller command to virtually shift impedance, thereby enhancing PHIL stability and extending its operational range without deteriorating the accuracy.

In addition, as presented in [13], the stability performance of the VSI and FBF methods is evaluated by analyzing the frequency responses of the open-loop transfer function of the PHIL setup for a range of example impedances. This analysis offers valuable insight into their comparative effectiveness and highlights the superior stability performance of the proposed VSI method over the FBF method. Furthermore, the applicability and effectiveness of the proposed VSI algorithm are validated using the PHIL setup shown in Fig. 8, where the VSI is implemented within a Triphase 90 kVA PA. Various IRs between MOI and HOI are employed to evaluate the stability performance of both VSI-based and FBF-based PHIL setup. The experimental results presented in [13] highlight the limitations of the FBF method in stabilizing the PHIL setup due to the constraints imposed by a low cut-off frequency. In contrast, for the same PHIL setup, the VSI-based interface maintains stability without compromising accuracy. This demonstrates that the proposed VSI algorithm offers a significant advantage over the commonly adopted FBF method. Furthermore, this demonstrates the capability of the proposed VSI approach to extend the range of PHIL setups with guaranteed stability and accuracy, facilitating the validation of novel smart energy technologies and controls across a broader range of scenarios.

2.2 Approaches for PHIL Accuracy Enhancement

To mitigate the impact of the time delay as identified in Sect. 1.2 on PHIL closed-loop power signal synchronisation accuracy, the following PHIL accuracy enhancement schemes have been developed in ERIGrid 2.0.

Sliding Discrete Fourier Transform (SDFT)-based delay compensation in the abc-frame: A SDFT-based time delay compensation scheme was developed in [5], leveraging its linear phase and sample-by-sample processing mechanism. As shown in Fig. 9, SDFT comprises a comb filter in series with multiple complex resonators that present liner phase and steep cut-off amplitude characteristics, allowing selective update of signal phase and amplitude over frequencies of interest without magnitude distortion and phase shift. SDFT-based delay compensation was achieved by manipulating the PA reference signal in the abc frame and adding an additional phase shift to the output of resonators with the specific resonant frequency. This enables time delay compensation on a harmonic-by-harmonics basis without signal magnitude distortion.

Phase Locked Loop (PLL)-based delay compensation in the dq-frame: Building on PA controller modifications, a PLL-based time delay compensation method was developed in [1]. As shown in Fig. 10, compensation is achieved by manipulating the PA command signal and introducing an additional phase (δ_{comp}) to the HOI voltage angle (δ_{GFC}), extracted via PLL. The compensated HOI voltage angle (δ'_{GFC}) is further used to convert the current controller output (in dq frame) into a PA modulating

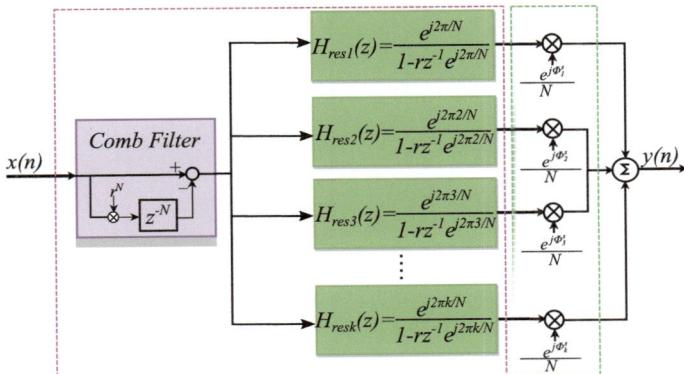

Fig. 9 SDFT-based PHIL delay compensation in abc-frame

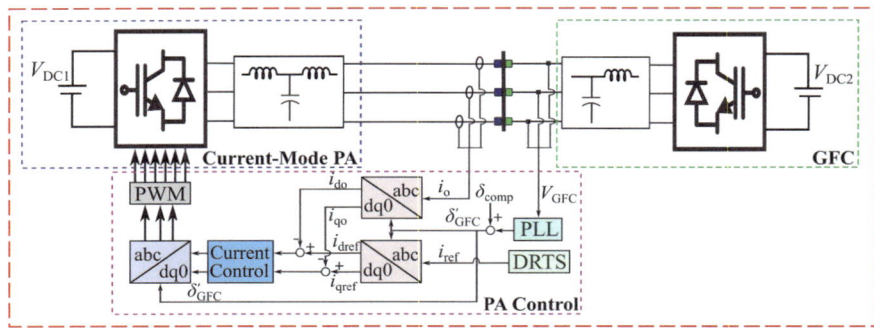

Fig. 10 PLL-based PHIL delay compensation in dq-frame

signal. This aligns the PA output current with its reference from MOI, effectively compensating for time delay and enabling accurate signal synchronization.

2.3 A Novel Framework for PHIL Sensitivity Analysis

To address the sensitivity issues from above and facilitate quantitative analysis and assessment of the impact of external disturbances on PHIL setups, a framework for sensitivity analysis of PHIL setups has been developed in [11]. A key contribution of this work is the analytical modeling of PHIL systems, focusing on disturbances (δ) causing sensitivity issues in two commonly used voltage-type and current-type ITM interfaces shown in Fig. 11. Accordingly, sensitivity transfer functions are derived for voltage-type (S_n^v) and current-type (S_n^i) ITM interfaces:

$$\begin{cases} S_1^v(s) = \dfrac{I_H(s)}{\delta I_H(s)} = \dfrac{1}{1+T_O^v(s)}, \\ S_2^v(s) = \dfrac{I_H(s)}{\delta V_H(s)} = \dfrac{1/Z_H(s)}{1+T_O^v(s)}, \\ S_3^v(s) = \dfrac{I_H(s)}{\delta V_{S'}(s)} = \dfrac{C_v(s)/Z_H(s)}{1+T_O^v(s)}, \\ S_4^v(s) = \dfrac{I_H(s)}{\delta I_S(s)} = \dfrac{-C_v(s)Z_S(s)}{[1+T_O^v(s)]Z_H(s)} \end{cases} \quad \begin{cases} S_1^i(s) = \dfrac{V_H(s)}{\delta V_H(s)} = \dfrac{1}{1+T_O^i(s)}, \\ S_2^i(s) = \dfrac{V_{S'}(s)}{\delta V_H(s)} = \dfrac{P_i(s)}{1+T_O^i(s)}, \\ S_3^i(s) = \dfrac{I_H(s)}{\delta V_H(s)} = \dfrac{-P_i(s)/Z_S(s)}{1+T_O^i(s)}, \\ S_4^i(s) = \dfrac{I_H(s)}{\delta V_H(s)} = \dfrac{-C_i(s)P_i(s)}{[1+T_O^i(s)]Z_S(s)} \end{cases}$$

These sensitivity functions are crucial to assess the robustness and enhanced stability properties of PHIL setups and to facilitate the sensitivity analysis of PHIL systems to external disturbances. Furthermore, as depicted in [11], the proposed framework reveals the inherent relationship between stability, accuracy, and sensitivity functions, enabling accurate estimation of PHIL system properties before its final-stage deployment. In addition to the sensitivity analysis criteria, Signal-

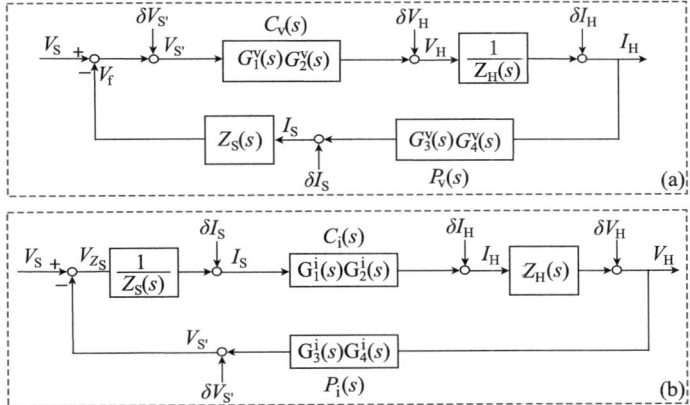

Fig. 11 PHIL diagrams with **a** voltage-type and **b** current-type ITM interfaces

to-Noise (SNR) and the Total Harmonic Distortion plus Noise (THD+N) metrics are employed to quantify the system sensitivity, facilitating the applicability of the proposed framework.

The applicability of the proposed sensitivity framework was validated through PHIL experiments in [11], demonstrating its effectiveness for testing both elementary and complex power system components. This framework serves as a guideline offering insights into PHIL design principles and sensitivity analysis.

3 Applications of HIL Techniques for CPES Testing

Aided by the advanced methods dedicated to improving the PHIL stability and accuracy, along with the sensitivity framework presented in Sect. 2, PHIL-based RTS techniques have been extensively leveraged to support the experimental validation of emerging smart energy techniques in the following test cases.

Case 1: Applying PHIL to support the development, calibration, and validation of digital twins of active distribution network: Digital Twins (DTs), serving as digital representations of physical assets within a network under various operational conditions, play a crucial role in facilitating the dynamic analysis of Distribution Networks (DNs). Although numerous DT approaches have been proposed in the literature, most have not been fully tested within network simulation models that accurately replicate real-world conditions. This limitation hinders a thorough assessment of their effectiveness in replicating the dynamic behavior of DNs. The joint research presented in [2] bridges this gap by conducting RTS within a PHIL experimental setup to support the development, calibration, and validation of measurement-based, data-driven, and reduced-order DTs for DNs. Figure 12 illustrates the PHIL experimental setup, which consists of a hybrid high-voltage and medium-voltage grid hosted on an

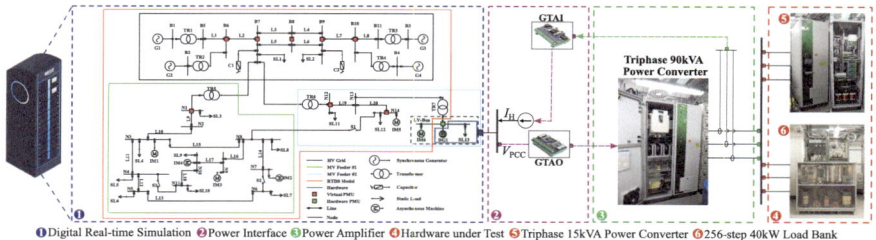

Fig. 12 Case 1—PHIL testing setup [2]

① Digital Real-time Simulation ② Power Interface ③ Power Amplifier ④ Hardware under Test ⑤ Triphase 15kVA Power Converter ⑥ 256-step 40kW Load Bank

RTS platform, while the actual laboratory represents the Low Voltage (LV) DN. Various DN configurations with a variety of load compositions, Distributed Generation (DG) penetration levels, and types of DG controls were implemented. Subsequently, dynamic responses to voltage variations, in terms of real and reactive power at the point of common coupling between the RTS system and LV DN, were collected to aid the development and calibration of the proposed DTs. Furthermore, the developed DTs were integrated into the RTS system to assess their effectiveness in digitally replicating physical LV DN under diverse scenarios. The results demonstrate that DTs accurately reproduce the dynamic real and reactive power responses of the LV DN to voltage disturbances, highlighting the effectiveness of the PHIL-aided DT development and validation approach.

Case 2: PHIL testing of grid forming control for power system black-start: As power grids transit toward high renewable penetration, Grid-forming Converter (GFC) is transforming black start strategies by facilitating renewable-rich power system restoration with fast response and enhanced resilience. A refined GFC Virtual Synchronous Machine (VSM) control was proposed in [1], tailored for black-start application. The VSM Q-V loop is adapted with voltage ramp-up to enable soft energization, effectively mitigating transformer inrush currents. In addition, the P-w and Q-V loops are modified to enhance voltage support for the restored AC network and enable smooth synchronization between restarted islands and the main grid. As shown in Fig. 13, a PHIL setup incorporating a hardware GFC is employed to validate the capability of the modified VSM controller to restart a simulated network hosted in the RTS platform. The current-type ITM interface method proposed in [3] is used to bridge hardware GFC with the RTS system. The stability and accuracy enhancement schemes, in particular the time delay compensation, presented in Sect. 2 were employed to facilitate a robust PHIL testing environment with enhanced accuracy. The proposed VSM control was evaluated under black-start sequences, encompassing soft transformer energization to mitigate inrush currents, critical load pickup, and grid synchronisation. The test results presented in [1, 3] not only demonstrate the effectiveness of the proposed GFC VSM control strategies but also enhance grid operators' confidence in adopting these control schemes for subsequent field tests within the Scottish power network. Furthermore, this PHIL configuration establishes a foundation for testing GFC control schemes tailored for black-start and grid ancil-

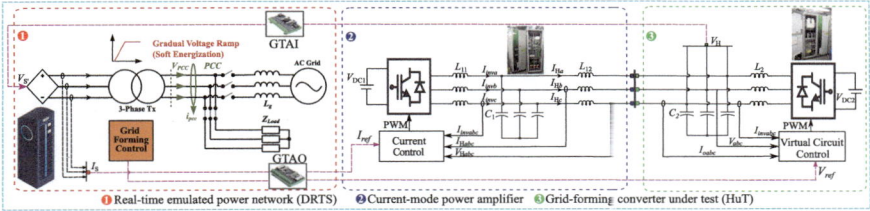

Fig. 13 Case 2—PHIL testing setup [1]

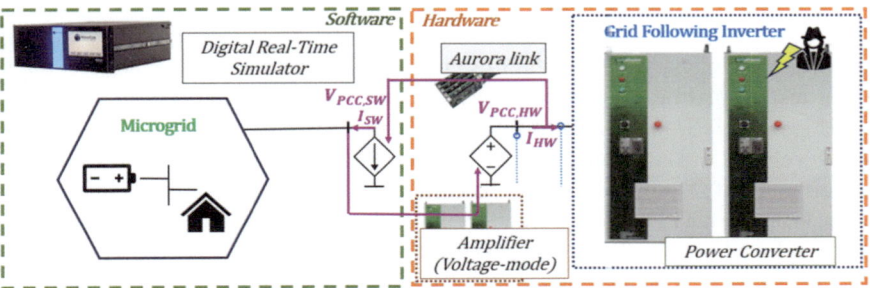

Fig. 14 Case 3—PHIL testing setup

lary services provision in a controlled environment across a wide range of network conditions.

Case 3: PHIL-based assessment of cyber resilience of inverter-dominated microgrids: In inverter-dominated Microgrids (MGs), new vulnerabilities arise due to the extensive use of control and communication technologies. Adversaries can intrude on the CPES, aiming to exploit their dynamics to cause stability and power quality issues. Research conducted in [9] explores the vulnerabilities and impact of stealthy cyber-attacks targeting the synchronization loops of Grid-forming (GFM) and Grid-following (GFL) inverters and their dynamic interactions. This theoretical analysis reveals that well-designed stealthy attacks can potentially lead to the disconnection of the inverter-interfaced resources or even an MG blackout. The PHIL experimental setup, presented in Fig. 14, comprises a real-time simulated MG and a hardware GFL inverter. This setup was leveraged to validate the analysis of the cyber resilience of inverter-dominated MG against stealthy cyber attacks on the synchronization loops of the inverters. It effectively demonstrated that appropriate design of the operational parameters of MG can mitigate the arising anomalies, enabling enhanced cyber resilience against such cyber attacks.

Case 4: PHIL for prototyping and testing digital twins of distributed energy resources: As depicted in case 1, DT acts as a digital replica of physical assets in energy systems, playing a vital role in enhancing system visibility for monitoring and control purposes. The development and validation of DTs necessitate a controlled, yet realistic, testing and validation environment to facilitate their adoption

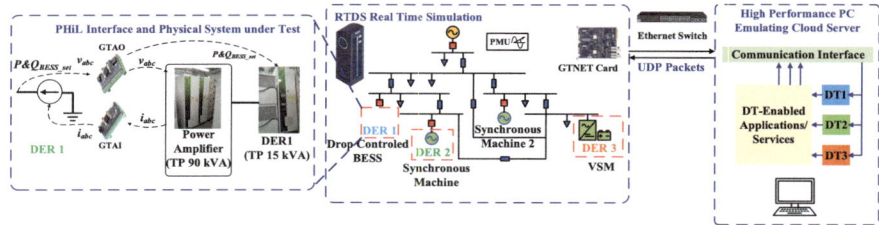

Fig. 15 Case 4—PHIL testing setup [7]

in the energy industry. To address this need, as illustrated in Fig. 15, an RTS-based PHIL testing platform dedicated to prototyping and testing the DTs of DERs was developed in [7, 8]. In this platform, a Visual Studio-based communication emulator was developed and integrated with RTS platform via Ethernet protocol to evaluate the performance of DTs for DERs under communication latency and timing jitter conditions. Moreover, a PHIL setup, aided by the compensation schemes presented above, was employed to integrate a physical GFL converter, replicating the DER within the RTS platform. Case studies were conducted in [7, 8] to validate the effectiveness of the proposed DT prototyping paradigm and testing platform in supporting DER-rich MG frequency control using DTs. With the increasing penetration of DERs in power systems, enhancing their coordination is essential, particularly by strengthening DER visibility despite limited communication constraints. From an application perspective, the improved visibility of the DER enabled by this DT prototyping and testing platform facilitates system-level monitoring and control applications, e.g. voltage and frequency regulation in DER-rich power grids. The proposed platform fosters the advancement of DTs for DERs and accelerates their adoption into modern energy systems.

Case 5: Advanced HIL testing chain for investigating interactions between CPES components during transients: In this case, the interactions between CPES components, particularly during transient power system events, are explored through an advanced HIL Testing Chain, as shown in Fig. 16. This Testing Chain is tailored for both academic and industrial users, such as relay and power electronics manufacturers, as well as distribution and transmission system operators. A range of HIL simulation setups are considered, from basic offline simulations to intricate PHIL simulations, thus creating a robust testing framework. As explained in [14], the different setups, e.g., RTS/Software-in-the-Loop (SIL), CHIL, PHIL and combined CHIL/PHIL, are suitable for investigating different dynamics of interest in modern power system studies. To demonstrate the applicability of this HIL Testing Chain, a case study is presented in [14], focusing on testing an advanced MG smooth mode transition algorithm. Particularly, the interactions between a sophisticated GFM inverter and existing GFL inverters under both normal and abnormal grid conditions are studied. By adhering to the steps of the proposed Testing Chain, valuable insights are gained into component validation during transient events.

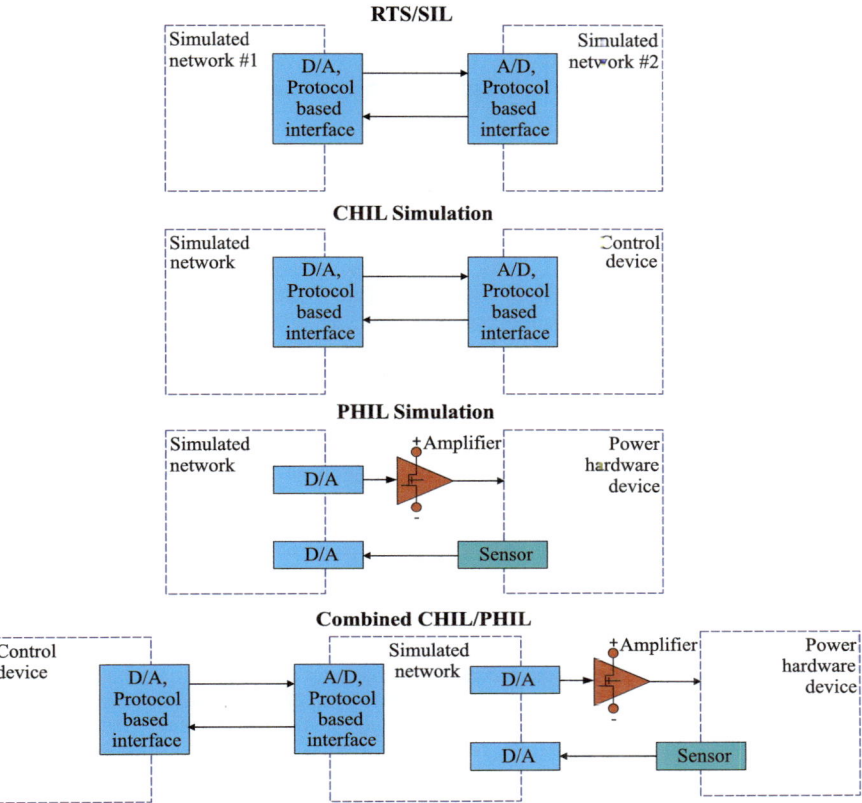

Fig. 16 Case 5—HIL simulation testing chain steps [14]

4 Closing Remarks

This chapter presents the improved RTS and HIL techniques developed in ERIGrid 2.0. Generic PHIL system modelling, along with the key research and application challenges regarding the stability, accuracy, and sensitivity analysis, are presented. Advanced methods developed to address these challenges are presented, including novel VSI and adaptive SP-aided stability enactment methods, SDFT and PLL-based time delay compensation for accuracy enhancement, and a novel PHIL sensitivity analysis framework.

These advancements have been employed to facilitate a variety of PHIL tests, including the development and validation of a digital twin for LV distribution networks, validation of an improved VSM control strategy for GFC-based black start testing, assessment of the inherent cyber resilience of inverter-dominated MGs against PLL attacks, prototyping and validation of digital twins for DER monitoring and control, and the implementation of an advanced HIL test chain designed to investigate interactions among CPES components. These advancements serve as key enablers for a more robust RTS and HIL testing with enhanced fidelity, facilitating the broader adoption of these techniques for smart power system validation.

References

1. Alassi A, Feng Z, Ahmed K, Syed M, Egea-Alvarez A, Foote C (2023) Grid-forming VSM control for black-start applications with experimental PHIL validation. Int J Electr Power Energy Syst 151:109119. https://doi.org/10.1016/j.ijepes.2023.109119
2. Barzegkar-Ntovom GA, Kontis EO, Papadopoulos TA, Feng Z, Burt GM (2024) Digital twin aided dynamic analysis of distribution networks with power hardware-in-the-loop validation. In: 2024 International conference on smart energy systems and technologies (SEST), pp 1–6. https://doi.org/10.1109/SEST61601.2024.10694671
3. Feng Z, Alassi A, Syed M, Pena-Alzola R, Ahmed K, Burt G (2022) Current-type power hardware-in-the-loop interface for black-start testing of grid-forming converter. In: IECON 2022 48th annual conference of the ieee industrial electronics society, pp 1–7. https://doi.org/10.1109/IECON49645.2022.9968517
4. Feng Z, Pena-Alzola R, Seisopoulos P, Guillo-Sansano E, Syed M, Norman P, Burt G (2020) A scheme to improve the stability and accuracy of power hardware-in-the-loop simulation. In: IECON 2020 The 46th annual conference of the IEEE industrial electronics society, pp 5027–5032. https://doi.org/10.1109/IECON43393.2020.9254407
5. Feng Z, Pena-Alzola R, Seisopoulos P, Syed M, Guillo-Sansano E, Norman P, Burt G (2021) Interface compensation for more accurate power transfer and signal synchronization within power hardware-in-the-loop simulation. In: IECON 2021 47th annual conference of the IEEE industrial electronics society, pp 1–8. https://doi.org/10.1109/IECON48115.2021.9589158
6. Feng Z, Pena-Alzola R, Syed MH, Norman PJ, Burt GM (2023) Adaptive Smith predictor for enhanced stability of power hardware-in-the-loop setups. IEEE Trans Ind Electron 70(10):10204–10214. https://doi.org/10.1109/TIE.2022.3224196
7. Han J, Hong Q, Feng Z, Burt G, Booth C (2023) Digital twins of distributed energy resources for real-time monitoring: data reporting rate considerations. In: IECON 2023- 49th annual conference of the IEEE industrial electronics society, pp 1–7. https://doi.org/10.1109/IECON51785.2023.10312555
8. Han J, Hong Q, Feng Z, Syed MH, Burt GM, Booth CD (2022) Design and implementation of a real-time hardware-in-the-loop platform for prototyping and testing digital twins of distributed energy resources. Energies 15(18). https://doi.org/10.3390/en15186629
9. Kontou A, Syed M, Paspatis A, Feng Z, Konstantinou C, Hatziargyriou N (2025) Exploiting the inherent cyber resilience of inverter-dominated microgrids against PLL attack. IEEE Trans Ind Electron, pp1–6. https://doi.org/10.1109/TIE.2025.3581270
10. Kotsampopoulos PC, Lehfuss F, Lauss GF, Bletterie B, Hatziargyriou ND (2015) The limitations of digital simulation and the advantages of PHIL testing in studying distributed generation provision of ancillary services. IEEE Trans Ind Electron 62(9):5502–5515. https://doi.org/10.1109/TIE.2015.2414899
11. Lauss G, Feng Z, Syed MH, Kontou A, Paola AD, Paspatis A, Kotsampopoulos P (2022) A framework for sensitivity analysis of real-time power hardware-in-the-loop (PHIL) Systems. IEEE Access 10:101305–101318. https://doi.org/10.1109/ACCESS.2022.3206780
12. Lauss G, Strunz K (2021) Accurate and stable hardware-in-the-loop (HIL) real-time simulation of integrated power electronics and power systems. IEEE Trans Power Electron 36(9):10920–10932. https://doi.org/10.1109/TPEL.2020.3040071
13. Paspatis A, Kontou A, Feng Z, Syed M, Lauss G, Burt G, Kotsampopoulos P, Hatziargyriou N (2024) Virtual shifting impedance method for extended range high-fidelity PHIL testing. IEEE Trans Ind Electron 71(3):2903–2913. https://doi.org/10.1109/TIE.2023.3269467

14. Paspatis A, Kontou A, Kotsampopoulos P, Lagos D, Vassilakis A, Hatziargyriou N (2024) Advanced hardware-in-the-loop testing chain for investigating interactions between smart grid components during transients. Electr Power Syst Res 228:109990. https://doi.org/10.1016/j.epsr.2023.109990
15. Ren W, Steurer M, Baldwin TL (2008) Improve the stability and the accuracy of power hardware-in-the-loop simulation by selecting appropriate interface algorithms. IEEE Trans Ind Appl 44(4):1286–1294. https://doi.org/10.1109/TIA.2008.926240

Open Access This chapter is licensed under the terms of the Creative Commons Attribution 4.0 International License (http://creativecommons.org/licenses/by/4.0/), which permits use, sharing, adaptation, distribution and reproduction in any medium or format, as long as you give appropriate credit to the original author(s) and the source, provide a link to the Creative Commons license and indicate if changes were made.

The images or other third party material in this chapter are included in the chapter's Creative Commons license, unless indicated otherwise in a credit line to the material. If material is not included in the chapter's Creative Commons license and your intended use is not permitted by statutory regulation or exceeds the permitted use, you will need to obtain permission directly from the copyright holder.

Laboratory Infrastructure Integration and Automation

A. Acosta, G. Silano, G. Paludetto, O. Gehrke, M. C. Pham, Q. T. Tran, V. Rajkumar, S. Vogel, and A. Monti

Abstract This chapter presents a software suite to support the implementation of distributed Research Infrastructure experiments. These tools provide communication interoperability and support Configuration Management. Moreover, we propose a laboratory middleware that extends these functionalities with the concept of Research Infrastructure as Code.

A. Acosta (✉) · A. Monti
RWTH Aachen University, Aachen, Germany
e-mail: andres.acosta@eonerc.rwth-aachen.de

A. Monti
e-mail: amonti@eonerc.rwth-aachen.de

G. Silano · G. Paludetto
Ricerca sul Sistema Energetico - RSE S.p.A, Milan, Italy
e-mail: giuseppe.silano@rse-web.it

G. Paludetto
e-mail: gabriele.paludetto@rse-web.it

O. Gehrke
Technical University of Denmark, Kgs. Lyngby, Denmark
e-mail: olge@dtu.dk

M. C. Pham · Q. T. Tran
French Alternative Energies and Atomic Energy Commission, Paris, France
e-mail: minh-cong.pham@cea.fr

Q. T. Tran
e-mail: quoctuan.tran@cea.fr

V. Rajkumar
Delft University of Technology, and TenneT TSO B.V, Delft, The Netherlands
e-mail: v.rajkumar@tudelft.nl

S. Vogel
OPAL-RT Germany, Nürnberg, Germany
e-mail: steffen.vogel@opal-rt.com

1 Large-Scale Integration of Laboratories

The work presented in this chapter builds on previous experience with experiments involving multiple RTIs in a variety of different configurations and across organisational boundaries [11]. This includes the real-time coupling of multiple geographically distributed physical laboratories as well as the coupling between physical laboratories and different types of simulators running in real-time [10], primarily in CHIL and PHIL configurations (see in chapter "Improved Hardware-in-the-Loop-based Testing").

One of the main insights gained in the process is that there are both technical and fundamentally non-technical obstacles to RTI integration. The former relate to the mechanics of moving data and/or signals from A to B, through firewalls and across large distances (data transport), while the latter are primarily artefacts of the meeting of different organisations, procedures and their underlying assumptions. Usually, the development focus tends to be on the former, leading to an ad-hoc, as-needed approach, which yields quick early results but soon becomes inefficient without addressing the latter.

One of the specific challenges of the CPES field is the large range of possible answers to the question "What is a Smart Energy testing laboratory?". Of special importance to RTI integration is the broad range of automation levels across RTIs and the large variety of approaches to laboratory automation. Therefore, this chapter presents several technical solutions to overcome these issues.

2 Preliminaries and Definitions

RTIs often focus on research and testing of CPES applications, feature tools and equipment for offline simulation, real-time simulation, and HIL. Moreover, communications are emulated, for instance, for testing CPES applications [1]. Consequently, RTIs often require defining tools and methods to provide interoperability, interconnect equipment from different vendors, and support multi-protocol communications. This is achieved through a combination of SCADA systems and custom solutions, e.g., those based on REST APIs. Several technological advancements led to the idea of geographically distributed interconnection of laboratories, which allows resource, more complex scenarios, and multidisciplinary collaboration. However, when the interconnection takes place over the public internet, there are challenges to keeping real-time constraints and providing cybersecurity, among others. Therefore, in RTI integration scenarios, we can distinguish between two types of communication:

- *Internal Communications*: As mentioned above, this communication takes place inside the RTI. It interconnects Digital Real-Time Simulator (DRTS) platforms, and CHIL and power equipment. Sometimes, the RTI can also feature a SCADA system and other tools for communication, such as APIs and custom scripts. Here,

Laboratory Infrastructure Integration and Automation

adapter refers to any module that provides interoperability between communication protocols at the local level. This concept will be refined later in Sect. 3.2.
- *Transport Communications*: Enable interconnection between geographically distributed RTIs. The underlying tools are called *Transports*. Several solutions providing the functionality of Transports have been implemented, taking into account different requirements. In the power and energy systems community, multiple efforts have resulted in tools such as HELICS [3], the Distributed Co-simulation Protocol [4], and OpSim [13]. ERIGrid 2.0 focuses on VILLASnode [6], JaNDER [7], and AIT's Lablink [9] frameworks.

As the number of devices and interconnections in a given RTI increases, the organization and configuration of laboratory equipment become more difficult. Aspects like a unified and automated naming convention for signal exchange, signal units and their scaling, and the way internal communications are implemented become more relevant as the number of participating RTIs increases in the distributed experiment [12]. This led to the idea of defining a set of tools to simplify the RTI integration process with the ability to enhance and extend the capabilities of the existing transport tools. These tools are presented below.

3 Tools for Research Infrastructure Integration

The two main types of communications involved in a multi-RI experiment (i.e., internal and transport communications) are outlined above. This section introduces three available transport tools and the universal API (uAPI) for internal communications. Moreover, the uAPI enables interoperability between transports and some of its main functionalities are shown through tests.

3.1 Laboratory Coupling Tools

The main purpose of these tools is to provide transport communications. As mentioned above, different projects and initiatives led to the implementation of multiple lab coupling tools, each of them with its advantages and disadvantages. The ERIGrid 2.0 project had the ambitious goal of demonstrating the interoperability of multiple lab coupling tools in distributed RTI experiments. In particular, three transport tools, which are described as follows, are considered:

- *VILLASnode* [2] is a flexible high-performance gateway supporting more than 20 communication protocols. VILLASnode is part of VILLASframework, a toolset to enable the interconnection of laboratory equipment. It covers a wide set of applications, ranging from local real-time co-simulations with dynamics in the range of microseconds to geographically distributed PHIL experiments.

- *JaNDER* [7] serves as a middleware solution designed to facilitate seamless data exchange between smart systems through secure, standardized APIs. Developed within the predecessor ERIGrid [10], JaNDER utilizes HTTPS for secure communication and a cloud-based Redis database[1] to replicate data between local smart systems and a central cloud node. Its core functionality involves mirroring infrastructure data bidirectionally—data recorded in a local database is simultaneously updated in the cloud and vice versa—ensuring interoperability across different transport services.
- *Lablink* [9] is a platform developed by AIT that provides communication between DRTS and HIL devices. The Lablink Core middleware enables communication between distributed clients, while communications are implemented using the MQTT protocol, which allows asynchronous messaging and supports secure connections. Synchronous communications are achieved using remote procedure calls. Moreover, the Lablink Core exposes interfaces that can be used by clients to interconnect hardware and simulators and by utilities to provide services on top, like synchronization.

3.2 Universal API

The uAPI was proposed as a unified protocol to implement internal communications and allow interoperability between transport platforms. Thus, RTI SCADA systems and other internal communication mechanisms can interconnect to remote RIs using any transport tool, provided that it supports uAPI. Besides, the uAPI provides harmonized functionality by defining a set of common methods to access the list of available signals, RIs, status, etc.

Being a REST API, the uAPI requires a server side and a client side. The server side is implemented as part of the transport tool, whereas the client side is implemented as an adapter module, which connects to the internal SCADA system or directly to the laboratory equipment. It is worth stressing that this concept enables a substantial increase in software reusability since the interoperability provided by the uAPI allows transparent interchange adapters between RTIs. In this way, once a new adapter is implemented to support a specific communication protocol, it becomes available immediately for the three transport tools.

The uAPI is structured according to the OpenAPI specification,[2] which aims to standardize an HTTP API. This specification outlines a set of guidelines for defining APIs, including the available endpoints, the data models employed, authentication mechanisms, and more, so that clients can understand how to interact with the API, as well as to automatically generate code and documentation. A major advantage of adhering to the OpenAPI specification is that it promotes uniformity and minimizes

---

[1] https://redis.io/.
[2] https://www.openapis.org/.

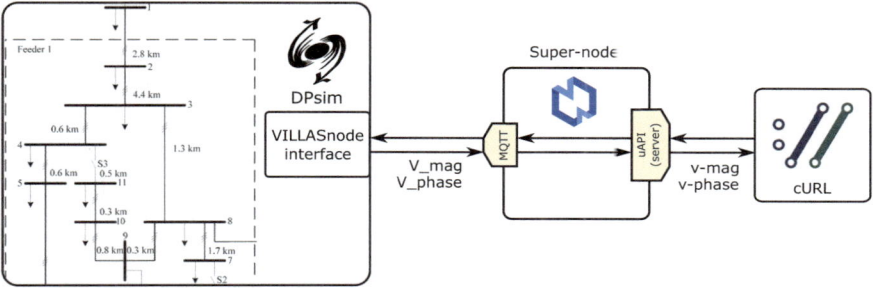

Fig. 1 Example setup for testing the uAPI.

the need for manual documentation. Offering a standardized method to describe API, it ensures that they are well-documented and easy to understand. Moreover, the OpenAPI specification facilitates the automatic generation of client libraries, significantly reducing the time developers need to invest. All documentation related to the developed uAPI can be accessed publicly through the GitHub Repository.[3]

To illustrate the uAPI's implementation, consider the setup shown in Fig. 1. Here, the open-source RTS DPsim [5] runs the CIGRE-MV benchmark. Node voltages are sent through the VILLASnode interface via MQTT to an instance of VILLAS node, which implements the server side of the uAPI. In this case, the uAPI functions using the cURL command line utility[4] can be verified. The interested reader can find an open-source implementation of this scenario in the corresponding GitHub Repository.[5]

In this case, VILLASnode runs on the local host machine. Thus, for this specific transport, the root endpoint of the uAPI corresponds to the URL http://localhost:8080/api/v2/universal/node_uapi/, where `node_uapi` is the name of the node defined in the VILLASnode configuration file. The cURL command to get a list of the available channels would be:

```
curl -v
   ↪ http://localhost:8080/api/v2/universal/node_api/channels
```

The channels are the signals that are sent from the DPsim simulator; in this case, a total of 30 signals is obtained, corresponding to the magnitudes and phases of the 15 buses found in the CIGRE-MV benchmark. An extract of the response from the uAPI server is shown in Fig. 2.

---

[3] https://erigrid2.github.io/JRA-3.1-api/universal-api.html.

[4] https://curl.se/.

[5] https://github.com/ERIGrid2/MOOC4.git.

Fig. 2 uAPI response for GET/channels

```
[
    {
        "id": "n1-v",
        "datatype": "float",
        "readable": true,
        "writable": false,
        "description": "Voltage N1",
        "rate": 1.0,
        "payload": "samples"
    },
    {
        "id": "n1-ph",
        "datatype": "float",
        "readable": true,
        "writable": false,
        "description": "Phase N1",
        "rate": 1.0,
        "payload": "samples"
    }
]
```

Fig. 3 uAPI response for GET/channel/sample

```
{
    "timestamp": 1742461686.0433388,
    "value": -0.029712912935014171,
    "validity": "unknown",
    "source": "unknown",
    "timesource": "unknown"
}
```

To get the value of a specific channel, the endpoint would be `channel/<channel-id>/sample`, being `<channel-id>` the id of the channel identified in the previous step, as shown in the following cURL request.

```
curl -v
    ↪ http://localhost:8080/api/v2/universal/node_uapi/channel/n15-ph/sample
```

which results in an output similar to the one in Fig. 3.

3.3 Configuration Management

The work in ERIGrid 2.0 builds on the foundation laid by its predecessor, ERIGrid, which pioneered experiments involving multiple RTIs, known as "multi-RI experiments," in various configurations [11]. These experiments included the real-time coupling of geographically distributed physical laboratories, as well as the integra-

Laboratory Infrastructure Integration and Automation 71

tion of physical laboratories with different types of simulators operating in real-time, particularly in PHIL and CHIL setups. During the demonstration phase of ERIGrid, valuable experience was gained in implementing closed-loop interconnections of RTIs across organizational boundaries. One of the key insights from this process was that many of the challenges encountered were fundamentally non-technical, arising primarily from the convergence of different organizations, procedures, and their underlying assumptions. The RTI interconnection concept was successfully demonstrated in ERIGrid, but no structured process was developed for cross-organizational experimental setups. Ad hoc adjustments were made, leading to challenges exacerbated by coordinating laboratory time between organizations. Key obstacles included the following points:

- A technical solution for signal exchange between RTIs was developed, but aligning signal interpretation across RIs required significant effort. Confusion over signal labels, units, conventions, scaling, and value ranges caused extensive debugging, repeated experiments, and delays.
- Repeating or resuming multi-RTI experiments was more complex than in a single lab, as it required recreating identical conditions across multiple labs with numerous DERs which are often reconfigured by other users, and added significant challenges.
- There are two main ways to record data in geographically distributed experiments: sending all data to a central logger in real-time or recording data locally at each site. The first approach often faces technical limits, such as bandwidth constraints, while the second requires merging logs from different sites. Beyond ensuring synchronized clocks, challenges arise if data labels and units are not standardized, mirroring the signal configuration issues described earlier.

While none of these obstacles are insurmountable or unsolvable, the lack of a systematic approach has led to significant manual, error-prone, and time-consuming efforts. These challenges are further amplified by the geographical distance between participants. Therefore, the objective of applying the Configuration Management (CM) approach in ERIGrid 2.0 is to automate the configuration of data exchange for multi-RTI experiments. This involves the following:

- *Offline Global Configuration/Description*: Defining the general data flow between the participating RTIs in an offline setting.
- *Online Automated Mapping*: Dynamically mapping local RTI signals to the appropriate data channels for the experiment, based on the global configuration.

The proposed CM approach is illustrated in Fig. 4, depicting two example RTIs conducting a joint experiment. Data exchange between these RTIs is enabled through dedicated laboratory coupling tools, such as the VILLAS Framework, JaNDER, or Lablink. The CM framework consists of two components. First, a module that splits and maps a global configuration file definition to multiple local files containing the specific configuration of a given RI participating in the distributed experiment, i.e., the CM tool. The second component is the ERIGrid 2.0 launcher, which parses this

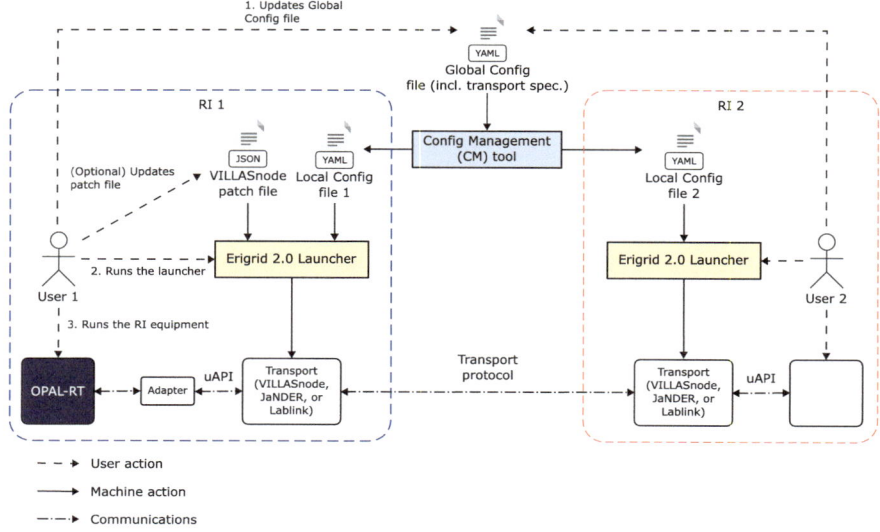

Fig. 4 Configuration management workflow and tools

local configuration file and runs the corresponding transport tool. The CM approach combines these components in a highly automated workflow comprising three steps:

1. The process starts with the definition of the shared global configuration file, where the transport communications and the input/output connections among RTIs are defined.
2. The user runs the ERIGrid 2.0 launcher, specifying the local configuration file produced by the CM tool as input.
3. Finally, the user starts the underlying TRI equipment, e.g., DRTS, grid emulators or load banks.

Although the process currently requires the user to run the CM tool, the level of automation can be increased through a Continuous Integration and Continuous Delivery/Deployment (CI/CD) task, through which the global configuration file is centrally managed using a version control system, e.g. Git. With every new commit/push of this file, a task is triggered to run the CM tool and new versions of the local configuration files are created. It is worth stressing that the approach does not disclose the way each RTI operates internally since only the configuration of the transport communications is shared. This implies that some transports like VILLASnode require an extra step to take care of configuring the internal communications. In the case of VILLASnode, the configuration is completed by adding a patch file as an input to the ERIGrid 2.0 launcher. Thus, the internal configuration of the RTIs can be kept locally in the patch file.

4 Towards a Laboratory Middleware

The tools described in the previous sections constitute an important step towards a unified approach for research infrastructure integration and automation. However, as pointed out above, there are still some configuration steps not covered by the CM tool, which require the lab operator to prepare, control (starting and stopping) and monitor the DRTS and hardware equipment. This is more problematic as the number of devices increases not only at the local level but also for distributed RTI experiments. Here it is proposed the development of a laboratory middleware that allows to manage RTI equipment and incorporates the tools described in the previous sections.

The primary purpose of the laboratory middleware is to enable seamless data exchange between Smart Systems. This middleware supports a range of systems, including physical laboratories, real-time and non-real-time simulators, and builds upon the ERIGrid 2.0 Transports. The main objectives of middleware are: (i) Allowing Smart Systems to switch effortlessly between transport tools (e.g., from JaNDER to VILLAS) without the need for multiple interfaces or adjustments to data formats and naming conventions, (ii) introducing desirable features not currently offered by existing transport modules through modular services, and (iii) centralising shared features from individual transport modules to avoid duplication. For example, a sophisticated time synchronization mechanism from one transport can be offered as a common service to benefit other transports with less advanced solutions.

The overall architecture consists of three layers, as shown in Fig. 5: (i) Infrastructure layer, (ii) platform layer, and (iii) software layer. The *infrastructure layer* is the foundational layer comprising physical lab setups, hardware devices, and SCADA systems. Through this layer, laboratory operators will be able to monitor and control multiple (distributed) devices from a central interface. This idea is supported by the fact that many DRTS and hardware devices can be controlled using APIs. The *platform layer* provides essential services, such as network emulation, time synchronization, and resource provision. Finally, the *software layer* facilitates data exchange through tools like VILLAS, JaNDER, and Lablink, with the uAPI offering a standardised set of functionalities. This layered approach ensures efficient, flexible, and unified communication across diverse Smart Systems.

An important building block of the laboratory middleware is RIasC, short for Research Infrastructure as Code. RIasC is a framework designed to accelerate distributed RTI experiments. It consists of a suite of tools and services that can be used independently or in combination to support these experiments. Infrastructure as Code (IasC) refers to the practice of managing and provisioning data centers using machine-readable definition files rather than traditional methods such as manual configuration or interactive tools. This approach manages both physical components, like bare-metal servers and virtual machines, along with their associated configurations. Typically, these definitions are stored in version control systems, and IasC can be implemented using either scripts or declarative definitions. The term is most commonly associated with promoting declarative methods for infrastruc-

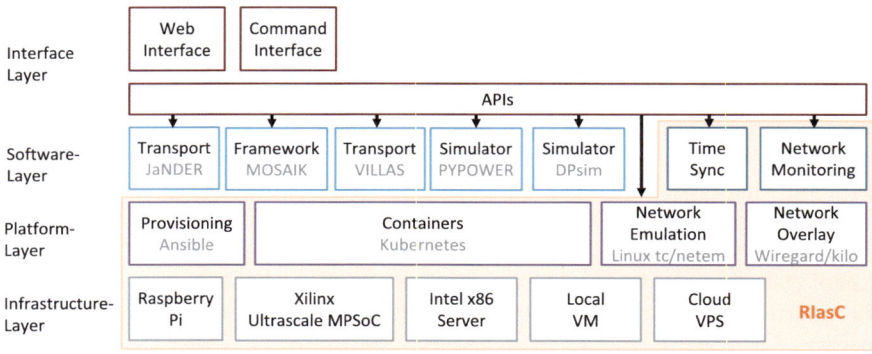

Fig. 5 Overview of the ERIGrid 2.0 laboratory middleware architecture [8]

ture management. RIasC realizes the principles of IasC by leveraging existing cloud computing technologies and applying them in research environments. RIasC offers several functionalities aimed at accelerating distributed RI experiments:

- *Provisioning of Mobile Units*: In RIasC, a mobile unit is a gateway that connects local laboratory equipment to remote laboratories. RIasC handles the setup, configuration, and maintenance of these units.
- *Deployment of containerised Software Components*: Mobile units join a Kubernetes cluster,[6] which allows for the declarative deployment of containerized applications.
- *Transparent Inter-Laboratory Overlay Network*: RIasC simplifies the process of setting up distributed experiments across multiple laboratories by providing a transparent IP overlay network that connects all participating labs.
- *Time Synchronization*: For geographically distributed coupled subsystems, RIasC provides a time synchronization service to ensure that experiments are executed with a common time base, allowing proper temporal alignment of simulation results, such as merged logs.
- *Network Emulation*: RIasC includes a network emulation service that allows researchers to define network characteristics (e.g., communication delay, packet loss, throughput) declaratively, enabling the realistic emulation of real-world network conditions.

The code base of RIasC has been released as open-source under the Apache 2.0 license. In addition, it includes a user-friendly website with useful documentation.[7]

---

[6] https://kubernetes.io.
[7] https://erigrid2.github.io/riasc/.

5 Final Thoughts

This chapter identified the main challenges faced by the implementation of multi-RTI experiments, with a focus on integration and automation. A suite of tools was proposed to harmonize communications at the RTI and the inter-RTI level and to facilitate configuration management. Concretely, the uAPI provides a uniform interoperability layer with useful data structures and functionalities. In addition, the CM tool enables a highly automated workflow to prepare experiments and run the transport tools at each RTI, with which RTIs can communicate. Finally, further automation necessities were identified, and a middleware architecture was presented, along with the RIasC concept.

References

1. Abdelrahman MS, Kharchouf I, Nguyen TL, Mohammed OA (2023) A hybrid physical co-simulation smart grid testbed for testing and impact analysis of cyber-attacks on power systems: framework and attack scenarios. Energies 16(23):7771. https://doi.org/10.3390/en16237771. Number: 23 Publisher: Multidisciplinary Digital Publishing Institute
2. Bach A, Monti A (2025) Remote real-time testing of physical components using communication setup automation. IEEE Access 13:39066–39075. https://doi.org/10.1109/ACCESS.2025.3546311. Conference Name: IEEE Access
3. Hardy TD, Palmintier B, Top PL, Krishnamurthy D, Fuller JC (2024) HELICS: a co-simulation framework for scalable multi-domain modeling and analysis. IEEE Access 12:24325–24347 (2024). https://doi.org/10.1109/ACCESS.2024.3363615. Conference Name: IEEE Access
4. Krammer M, Schuch K, Kater C, Alekeish K, Blochwitz T, Materne S, Soppa A, Benedikt M (2019) Standardized integration of real-time and non-real-time systems: the distributed co-simulation protocol. In: Modelica, pp 157–009. https://2019.international.conference.modelica.org/proceedings/html/papers/Modelica2019paper1C3.pdf
5. Mirz M, Dinkelbach J, Monti A (2020) DPsim-Advancements in power electronics modelling using shifted frequency analysis and in real-time simulation capability by parallelization. Energies 13(15):3879. https://doi.org/10.3390/en13153879. Publisher: Multidisciplinary Digital Publishing Institute
6. Monti A, Stevic M, Vogel S, Doncker RW, Bompard E, Estebsari A, Profumo F, Hovsapian R, Mohanpurkar M, Flicker JD, Gevorgian V, Suryanarayanan S, Srivastava AK, Benigni A (2018) A global real-time superlab: enabling high penetration of power electronics in the electric grid. IEEE Power Electron Mag 5(3):35–44. https://doi.org/10.1109/MPEL.2018.2850698
7. Pellegrino L, Pala D, Bionda E, Rajkumar VS, Bhandia R, Syed MH, Guillo-Sansano E, Jimeno J, Merino J, Lagos D, Maniatopoulos M, Kotsampopoulos P, Akroud N, Gehrke O, Heussen K, Tran QT, Nguyen VH (2020) Laboratory coupling approach, pp 67–86. Springer International Publishing. https://doi.org/10.1007/978-3-030-42274-5_5
8. Rajkumar V, Silano G, Gehrke O, Vogel S, Widl E, Paludetto G, Rikos E, Zerihun TA, Stefanov A, Palensky P, Strasser TI (2024) Laboratory middleware for the cyber-physical integration of energy research infrastructures. In: 2024 12th workshop on modeling and simulation of cyber-physical energy systems (MSCPES), pp 1–5. https://doi.org/10.1109/MSCPES62135.2024.10542755
9. Stahleder D, Reihs D, Lehfuss F (2018) LabLink-A novel co-simulation tool for the evaluation of large scale EV penetration focusing on local energy communities. In: CIRED 2018 Ljubljana workshop on microgrids and local energy communities. AIM, Ljubljana, Slovenia. Publisher: AIM

10. Strasser T, de Jong E, Sosnina M (2020) European guide to power system testing: the ERIGrid Holistic approach for evaluating complex smart grid configurations. Springer. https://doi.org/10.1007/978-3-030-42274-5
11. Strasser TI, Andren FP, Widl E et al (2018) An integrated pan-European research infrastructure for validating smart grid systems. e & i Elektrotechnik und Informationstechnik 135(8):616–622. https://doi.org/10.1007/s00502-018-0667-7
12. Strasser TI, Moyo C, Bründlinger R, Lehnhoff S, Blank M, Palensky P, van der Meer AA, Heussen K, Gehrke O, Rodriguez JE, Merino J, Sandroni C, Verga M, Calin M, Khavari A, Sosnina M, de Jong E, Rohjans S, Kulmala A, Mäki K, Brandl R, Coffele F, Burt GM, Kotsampopoulos P, Hatziargyriou N (2017) An integrated research infrastructure for validating cyber-physical energy systems. In: Marík V, Wahlster W, Strasser T, Kadera P (eds) Industrial applications of holonic and multi-agent systems. Springer International Publishing, Cham, pp 157–170. https://doi.org/10.1007/978-3-319-64635-0_12
13. Vogt M, Marten F, Montoya J, Töbermann C, Braun M (2019) A REST based co-simulation interface for distributed simulations. In: 2019 IEEE Milan PowerTech, pp 1–6. https://doi.org/10.1109/PTC.2019.8810661

Open Access This chapter is licensed under the terms of the Creative Commons Attribution 4.0 International License (http://creativecommons.org/licenses/by/4.0/), which permits use, sharing, adaptation, distribution and reproduction in any medium or format, as long as you give appropriate credit to the original author(s) and the source, provide a link to the Creative Commons license and indicate if changes were made.

The images or other third party material in this chapter are included in the chapter's Creative Commons license, unless indicated otherwise in a credit line to the material. If material is not included in the chapter's Creative Commons license and your intended use is not permitted by statutory regulation or exceeds the permitted use, you will need to obtain permission directly from the copyright holder.

Sector Coupling and Multi-Domain Systems Validation

E. Widl, G. Silano, O. Gehrke, and T. Zerihun

Abstract The transition to a decarbonized energy system requires integrating multiple energy carriers (e.g., power, heat, gas) and engineering domains (e.g., control, ICT) to optimise resource use and enhance resilience. The increasing share of non-programmable renewables, such as solar and wind, introduces grid stability challenges, including voltage fluctuations and congestion, necessitating advanced flexibility solutions. Multi-energy and multi-domain systems address these challenges. This chapter showcases innovative methods and approaches for assessing sector coupling and multi-energy systems in the context of the energy transition.

1 Technical Assessment of Integrated Energy Solutions

Sector coupling and multi-energy systems are expected to become crucial for the energy transition as they have the potential to enhance energy efficiency and integrate renewable energy sources more effectively by linking electricity, heating, cooling, and transport sectors [4]. This interconnected approach can leverage a more flexible use of energy, reducing reliance on fossil fuels and enabling higher shares of renewables, such as wind and solar. However, the assessments of the impact on the related technical infrastructure—such as the electrical grid, district heating system, and gas networks—is a relatively recent subject [3, 9]. For instance,

---

E. Widl (✉)
AIT Austrian Institute of Technology, Vienna, Austria
e-mail: edmund.widl@ait.ac.at

G. Silano
Ricerca sul Sistema Energetico - RSE S.p.A, Milan, Italy
e-mail: giuseppe.silano@rse-web.it

O. Gehrke
Technical University of Denmark, Kgs. Lyngby, Denmark
e-mail: olge@dtu.dk

T. Zerihun
SINTEF Energy Research, Trondheim, Norway
e-mail: tesfaye.zerihun@sintef.no

© The Author(s) 2025
T. I. Strasser et al. (eds.), *European Guide to Smart Energy System Testing*,
SpringerBriefs in Energy, https://doi.org/10.1007/978-3-031-99451-7_7

power-to-heat technologies, combined with thermal storage and flexible loads, enable surplus renewable electricity to be converted into thermal energy, supporting grid balancing and congestion management [5].

ERIGrid 2.0 has supported research on sector coupling and multi-energy systems by providing researchers with access to advanced laboratories and simulation tools for testing integrated energy solutions. This research and development performed in ERIGrid 2.0 (as described in chapters "Holistic Smart Energy System Validation"–"Laboratory Infrastructure Integration and Automation") has advanced innovative methods for assessing sector coupling and multi-energy systems in the context of the energy transition. This enables the validation of cross-sector energy applications—such as power-to-heat, hybrid energy networks, and smart control via ICT networks—ensuring efficient renewable integration and system flexibility.

The following sections provide examples of technical assessments of sector coupling and multi-energy systems. These examples cover different types of systems (multi-energy, cross-sector) and approaches (simulation, laboratory testing). Together, they demonstrate what insights may be gained using the tools and methodologies established in ERIGrid 2.0, showcasing the potential of integrated testing, validation, and simulation tools, enabling more efficient and reliable system integration for a broad spectrum of applications.

2 Co-Simulation of a Multi-energy Network

The Multi-Energy Networks benchmark (see section "Multi-Energy Network Benchmark") provides a reference setup for multi-energy sector coupling, where a power-to-heat facility connects a low-voltage electrical grid to a local heating network. The intention behind this benchmark scenario is to showcase the feasibility of using simulations for assessing multi-energy systems. However, simulation-based technical evaluations of multi-energy grids, primarily concentrating on operation and control, remain a challenge with the majority of available tools. This is because existing simulation tools generally focus on a single technical domain (e.g., thermal or electrical grids), having either emerged from extensive research in their particular scientific community or driven by a need from the industry to solve specific objectives. Hence, the co-simulation approaches developed as part of ERIGrid 2.0 (see section "Heat Networks and Power System") have been used to demonstrate the maturity of available simulation tools for the assessment of multi-energy grids.

Two implementations of the simulation benchmark have been created, relying on the mosaik co-simulation framework [6]. Both utilize pandapower [10] for simulating the electrical subsystem and employ independent implementations of the controllers' logic. Nonetheless, the implementations vary in their representation of the thermal domain. The first implementation uses the pandapipes package [2] for modeling the heat network, whereas the second implementation uses the DisHeatLib library [1] to simulate the complete thermal subsystem, including the heat network and the power-to-heat facility.

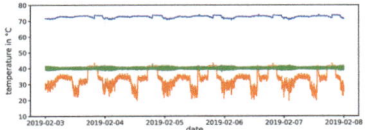 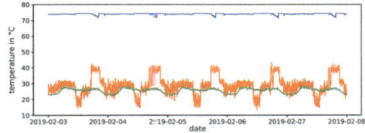

(a) Temperature profiles at supply line inlet (blue) and return line outlet (green) of *consumer_1* and network return line (orange).

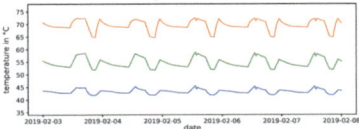 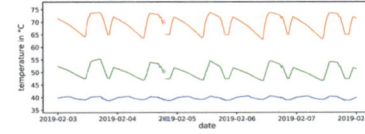

(b) Temperature profiles in storage tank from top (orange), center (green) and bottom (blue) of the volume.

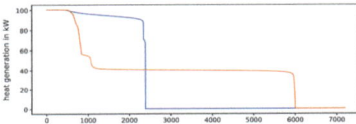 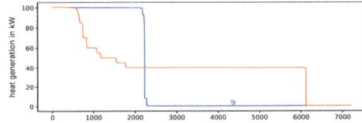

(c) Duration plot of heat pump power consumption with (orange) and without (blue) voltage control.

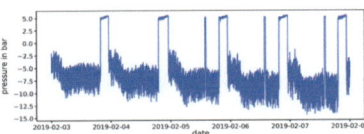 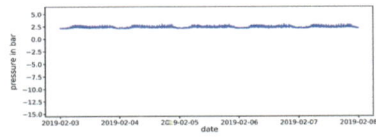

(d) Pressure profiles at return line outlet of *consumer_1*.

Fig. 1 Comparison of simulation benchmark results for pandapipes (left) and the DisHeatLib (right) [11]

Figure 1 shows a selection of results from the co-simulation setup using pandapipes (left column) and the DisHeatLib (right column), respectively [11]. Figure 1a compares the temperature profiles at the supply line inlet (blue) and the return line outlet (green) of *consumer_1* as well as the network return line (orange). This shows that both implementations differ considerably in how they model the consumer (i.e., the temperature of the returned mass flow), but the resulting network dynamics are comparable both qualitatively and quantitatively. The same argument holds for the operation of the storage tank (cf. Fig. 1b) and the heat pump (cf. Fig. 1c). In contrast, Fig. 1d shows that the computation of the pressure distribution fails with pandapipes in case more than one source feeds the network (i.e., when both the external thermal grid and the storage tank feed into the network's supply line). Wherever the pressure calculation does not fail, the comparison shows that the two approaches do not produce similar results due to different assumptions regarding the pressure drop in the consumer models.

This provides an interesting comparison between the distinct simulation approaches used by the two implementations. The pandapipes package provides a (quasi-)static analysis of balanced fluid systems, useful for the computation of temperature, pressure and velocity distributions in pipe networks. The DisHeatLib library is designed for the analysis of thermo-hydraulic transients in fluid systems, which is useful for the assessment of flow reversals and time-delayed propagation of fluid properties in pipe systems. Nevertheless, for the operation of the power-to-heat facility, both implementations give compatible results. This suggests that both approaches are suitable for adequately representing the dynamics of the coupled system and generating plausible results.

3 Multi-RTI Lab Tests of Multi-Energy District Flexibility

The Multi-Energy District Flexibility demonstration was designed to assess the feasibility and technical performance of power-to-heat technologies within a local multi-energy district. In contrast to the approach chosen in section "Heat Networks and Power System", this demonstration relied on laboratory infrastructure available within the ERIGrid 2.0 consortium. The demonstration focused on evaluating their impact on both the electrical and thermal networks, testing their ability to support grid operations through congestion management and balancing power provision. As part of ERIGrid 2.0, this experiment played a pivotal role in validating cross-domain energy system integration methodologies (see chapters "Holistic Smart Energy System Validation"–"Laboratory Infrastructure Integration and Automation"), leveraging geographically distributed RTIs and real-time control strategies to address key challenges of the energy transition.

The experiment was conducted across multiple RTIs, each contributing specific resources and expertise. As illustrated in Fig. 2, the setup included a district heating network at DTU, a CHP system at RSE, a heat pump with thermal storage at CRES, a BESS at SINTEF, and a real-time simulation of the distribution grid at TUD. The

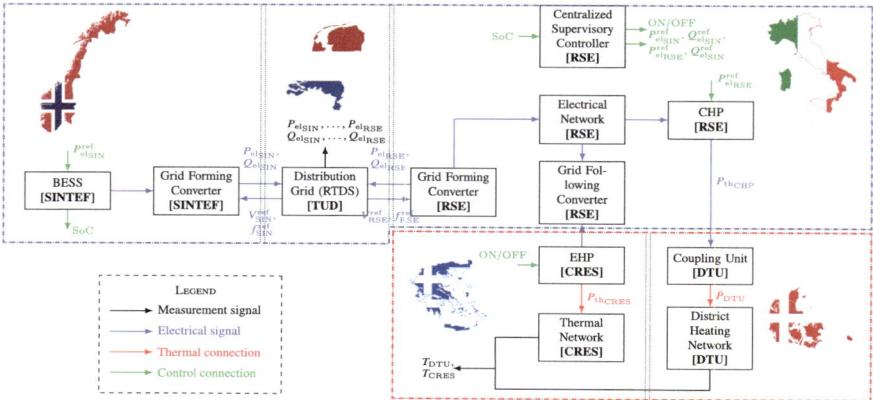

Fig. 2 Representation of the multi-energy district case study, with the electrical and thermal subsystems highlighted by blue and red dashed boxes, respectively [8]

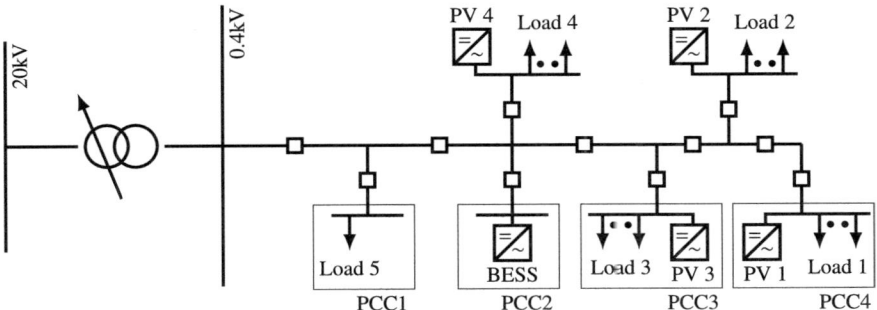

Fig. 3 Schematic representation of the CIGRE LV-distribution benchmark grid along with the PCCs [8]

electrical network was based on the CIGRE LV-distribution benchmark (see Fig. 3), emulated through DRTS, while the thermal system consisted of a double-pipe heating network interfaced with controllable heat sources and loads [7, 8].

The experimental infrastructure was designed to replicate the operation of a real multi-energy district by coupling electrical and thermal networks across geographically distributed RTIs. Data exchange was enabled through ERiGrid 2.0-developed tools, including JaNDER and uAPI (see chapter "Laboratory Infrastructure Integration and Automation"), ensuring reliable real-time communication between laboratories. As shown in Table 1, a diverse set of electrical and thermal parameters was exchanged among the RTIs, facilitating coordinated control of active and reactive power, voltage regulation, and thermal energy management. The integration of these technologies allowed for HIL testing and GDS, enabling the validation of advanced control strategies for multi-energy systems.

The demonstration focused on two key operational scenarios: Overvoltage management and undervoltage management. The first scenario, depicted in Fig. 4, addressed voltage rise due to high photovoltaic generation and low demand by increasing power consumption through heat pumps and BESS charging. The second scenario, shown in Fig. 5, examined voltage drops caused by low PV generation and high demand, requiring additional power generation from the CHP unit and thermal storage activation. The experimental results, illustrated in Figs. 6 and 7, confirmed

Table 1 Signals exchanged among RTIs, including their symbols and operational ranges [8]

Symbols	Unit	Min	Max	Symbols	Unit	Min	Max
$P_{el_{SIN}}$	kW	-40	40	$Q_{el_{SIN}}$	$kVAr$	-5	5
$P_{th_{CHP}}$	kW	46	81	$P_{el_{SIN}}^{ref}$	kW	-40	40
$\bar{P}_{DTU}$	kW	0	25	SoC	%	0	100
V_{SIN}^{ref}	V	150	400	f_{SIN}^{ref}	Hz	48	52
$P_{el_{RSE}}$	kW	-100	100	$Q_{el_{RSE}}$	$kVAr$	-50	50
V_{RSE}^{ref}	V	150	400	f_{RSE}^{ref}	Hz	48	52
$P_{th_{CRES}}$	kW	0	30	T_{DTU}	°C	0	100

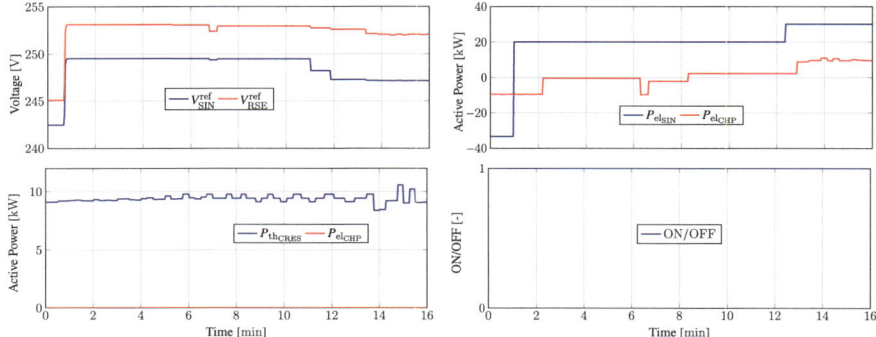

Fig. 4 Reference voltage sent from the distribution grid (TUD) to the grid-forming converters (SINTEF and RSE), along with the active power generated by the BESS and CHP and enable signals of the EHP in the overvoltage scenario [8]

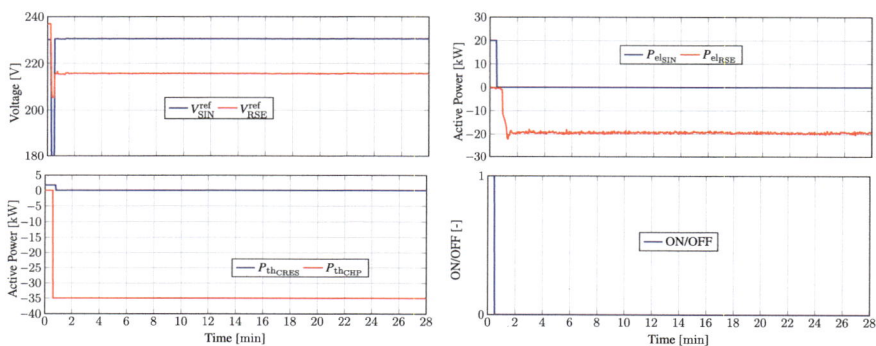

Fig. 5 Reference voltage sent from the distribution grid (TUD) to the converters (SINTEF and RSE), along with the active power generated by the BESS and CHP, and the enable signals of the EHP, in the undervoltage scenario [8]

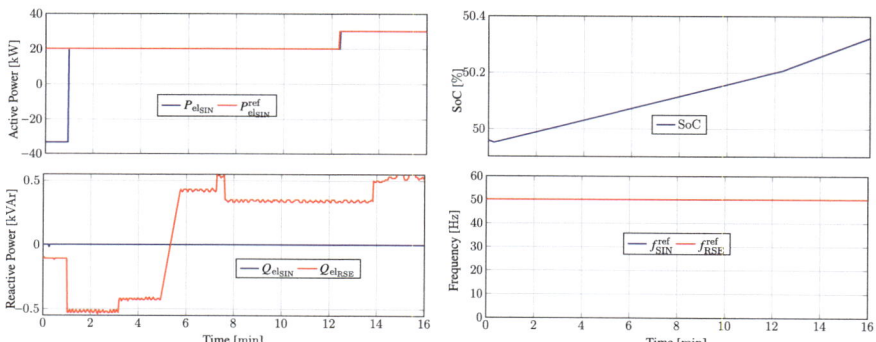

Fig. 6 Reference and actual active power profiles at the SINTEF facility, along with the SoC of the BESS, during the overvoltage scenario experiment. Additionally, reactive power and frequency data are included [8]

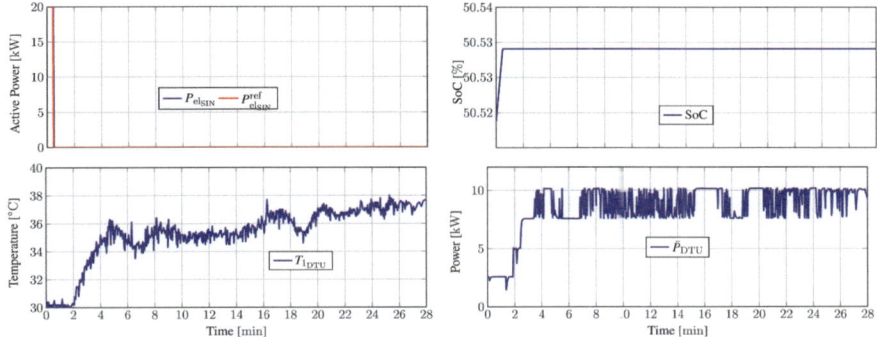

Fig. 7 Power and SoC of the BESS at SINTEF, alongside the temperature of the heat source buffer tank, and the power of heat sent to the buffer, in the undervoltage scenario [8]

that power-to-heat solutions effectively mitigated voltage deviations, demonstrating their potential as grid flexibility resources. Further details can be found in [7]. For the sake of brevity, plots and tabulated data are not included in this section.

The control strategy employed in the demonstration prioritized active power adjustments, while other parameters such as temperature control remained secondary considerations. The dynamic response of the thermal system, as shown in Fig. 8, highlighted the significantly slower time scales of thermal processes compared

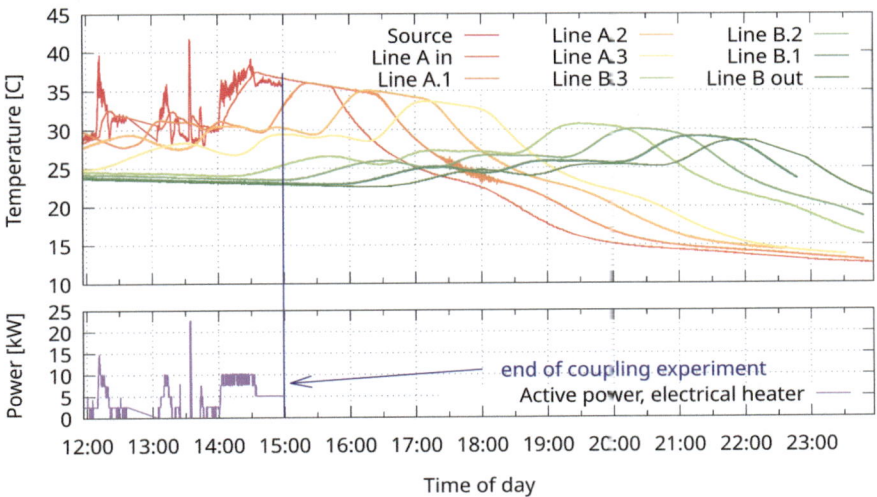

Fig. 8 Dynamic response of the thermal network. Bottom graph: Power consumption of the heat source, tracking the CHP's heat output. Top graph: Heat propagation through the network, measured at various points along the forward pipe [8]

to electrical systems, emphasizing the need for integrated control approaches that account for these differences.

By successfully integrating power-to-heat solutions within a distributed experimental framework, this demonstration validated the effectiveness of ERiGrid 2.0 methodologies, including the uAPI for RI data exchange, the JaNDER middleware for RTI communication (see chapter "Laboratory Infrastructure Integration and Automation"), and holistic test case descriptions for multi-energy system coordination (see chapters "Holistic Smart Energy System Validation"–"Enhanced Validation Methods and Benchmark"). The results confirmed that sector coupling can enhance grid flexibility, supporting both real-time operational stability and long-term energy transition goals. Furthermore, the use of advanced data exchange and synchronization tools facilitated seamless multi-laboratory collaboration, reinforcing the feasibility of cross-border experimental validation for emerging energy technologies.

4 Co-Simulation of an AC Grid and ICT Network

This demonstration was designed to characterize and validate co-simulation tools for multi-domain systems that combine an electrical grid and a networked ICT system, where the ICT infrastructure is used to monitor and control aspects of the electrical grid. In such systems, the performance of the ICT network directly affects the performance of the electrical grid, making it crucial that the modeling of the energy system accurately captures the interactions between the two subsystems. Currently, there are no established tool chains or validation approaches for comprehensively evaluating such cyber-physical power systems. To validate the proof-of-concept simulation developed within ERIGrid 2.0 (see section "Co-Simulation of Power Systems and ICT Networks"), a direct validation method was devised which compares the co-simulation results to experimental results from an emulated power-ICT system as well as from a physical laboratory.

The demonstration scenario considers the coordination between a Distribution System Operator (DSO) and an Aggregator for activating flexibility resources for voltage regulation. The test case assumes a section of an electrical distribution grid with two aggregators operating independently. Each aggregator manages a portfolio of flexible DERs which provide load reduction on demand. The primary customer for these grid services is the local DSO, which monitors the grid and submits service activation requests to the aggregators. A heterogeneous communication network facilitates data transfer between DSO, measurement equipment, flexibility assets, and aggregators. The test system is stimulated in two ways: through events within the communication network (equipment failure) or within the electrical network (sudden changes in load).

5 Summary and Implications

The access to state-of-the-art laboratories and simulation tools in ERIGrid 2.0 has enabled research and development dedicated to the evaluation of integrated energy solutions, enhancing the understanding of sector coupling and multi-domain systems. As a consequence of this research conducted as part of ERIGrid 2.0, new tools and methods have been developed for the validation of energy applications across multiple sectors. This is relevant for a variety of subjects of high importance for the energy transition–including power-to-heat, smart control, and hybrid energy networks–enabling the successful integration of renewable energy sources and improvement of system adaptability.

The sector coupling and multi-energy system examples presented here encompass a diverse set of systems and evaluation methods. Together they showcase the capabilities of ERIGrid 2.0's methodologies and tools for integrated testing, validation, and simulation, illustrating how these tools can enhance system integration across multiple domains.

References

1. Leitner B, Widl E et al (2019) A method for technical assessment of power-to-heat use cases to couple local district heating and electrical distribution grids. Energy 182:729–738. https://doi.org/10.1016/j.energy.2019.06.016
2. Lohmeier D, Cronbach D et al (2020) Pandapipes: an open-source piping grid calculation package for multi-energy grid simulations. Sustainability 12(23). https://doi.org/10.3390/su12239899
3. Lund H (2018) Renewable heating strategies and their consequences for storage and grid infrastructures comparing a smart grid to a smart energy systems approach. Energy 151:94–102. https://doi.org/10.1016/j.energy.2018.03.010
4. Mancarella P (2014) MES (multi-energy systems): an overview of concepts and evaluation models. Energy 65:1–17. https://doi.org/10.1016/j.energy.2013.10.041
5. Manco G, Tesio U, Guelpa E, Verda V (2024) A review on multi energy systems modelling and optimization. Appl Thermal Eng 236:121871. https://doi.org/10.1016/j.applthermaleng.2023.121871
6. Ofenloch A, Schwarz JS, Tolk D, Brandt T, Eilers R, Ramirez R, Raub T, Lehnhoff S (2022) MOSAIK 3.0: combining time stepped and discrete event simulation. In: 2022 open source modelling and simulation of energy systems (OSMSES). Aachen (2022)
7. Silano G, Paludetto G, Rodio C, Lazzari R, Feng Z, Kontou A, Paspatis A, Kotsampopoulos P, Gehrke O, Zerihun T, Acosta A, Rikos E, Subramaniam Rajkumar V (2024) D-JRA4.3 Demonstration of the extended research infrastructure. https://doi.org/10.5281/zenodo.12796352
8. Silano G, Rikos E, Rajkumar V, Gehrke O, Zerihun TA, Rodio C, Lazzari R (2024) Integrating power-to-heat services in geographically distributed multi-energy systems: a case study from the ERIGrid 2.0 project. In: 2024 open source modelling and simulation of energy systems (OSMSES), pp 1–6. https://doi.org/10.1109/OSMSES62085.2024.10668976
9. Sorknæs P, Lund H, Andersen AN (2015) Future power market and sustainable energy solutions—the treatment of uncertainties in the daily operation of combined heat and power plants. Appl Energy 144:129–138. https://doi.org/10.1016/j.apenergy.2015.02.041

10. Thurner L, Scheidler A et al (2018) Pandapower-An open-source python tool for convenient modeling, analysis, and optimization of electric power systems. IEEE Trans Power Syst 33(6):6510–6521. https://doi.org/10.1109/TPWRS.2018.2829021
11. Widl E, Wild C, Heussen K, Rikos E, Hoang TT (2022) Comparison of two approaches for modeling the thermal domain of multi-energy networks. In: 2022 open source modelling and simulation of energy systems (OSMSES), pp 1–6. https://doi.org/10.1109/OSMSES54027.2022.9769129

Open Access This chapter is licensed under the terms of the Creative Commons Attribution 4.0 International License (http://creativecommons.org/licenses/by/4.0/), which permits use, sharing, adaptation, distribution and reproduction in any medium or format, as long as you give appropriate credit to the original author(s) and the source, provide a link to the Creative Commons license and indicate if changes were made.

The images or other third party material in this chapter are included in the chapter's Creative Commons license, unless indicated otherwise in a credit line to the material. If material is not included in the chapter's Creative Commons license and your intended use is not permitted by statutory regulation or exceeds the permitted use, you will need to obtain permission directly from the copyright holder.

Experiences with Smart System Integration and Validation

G. Silano, A. Kontou, Z. Feng, A. Acosta, O. Gehrke, T. Zerihun,
S. Sanchez-Acevedo, G. Arnold, J. E. Rodriguez-Seco, R. Lazzari, C. Rodio,
and L. Pellegrino

1 Energy Vision and Smart Systems Role

Europe is charting a course towards a sustainable, secure, and competitive energy future with ambitious targets set for 2030[1] and 2050.[2] Central to these objectives is the integration of distributed renewable energy sources, which not only decarbonise the energy system but also modernise it to manage increasing complexity. The European Green Deal[3] and related policies underscore the need for a flexible, resilient, and intelligent energy ecosystem, transforming traditional power grids into sophisticated, cyber-physical networks.

---

[1] https://climate.ec.europa.eu/eu-action/climate-strategies-targets/2030-climate-targets_en.
[2] https://climate.ec.europa.eu/eu-action/climate-strategies-targets/2050-long-term-strategy_en.
[3] https://commission.europa.eu/strategy-and-policy/priorities-2019-2024/european-green-deal_en

---

G. Silano (✉) · R. Lazzari · C. Rodio · L. Pellegrino
Ricerca sul Sistema Energetico - RSE S.p.A, Milan, Italy
e-mail: giuseppe.silano@rse-web.it

R. Lazzari
e-mail: riccardo.lazzari@rse-web.it

C. Rodio
e-mail: carmine.rodio@rse-web.it

L. Pellegrino
e-mail: luigi.pellegrino@rse-web.it

A. Kontou
National Technical University of Athens, Athens, Greece
e-mail: alkistiskont@mail.ntua.gr

Z. Feng
University of Strathclyde, Glasgow, UK
e-mail: zhiwang.feng@strath.ac.uk

© The Author(s) 2025
T. I. Strasser et al. (eds.), *European Guide to Smart Energy System Testing*,
SpringerBriefs in Energy, https://doi.org/10.1007/978-3-031-99451-7_8

This energy transition introduces significant challenges. The intermittent nature of renewable sources, rapid advancements in digital technologies, and the emergence of controllable loads—such as electric vehicles, heat pumps, and energy storage systems—necessitate a complete rethinking of grid planning, operation and management. Moreover, market liberalisation and evolving regulatory frameworks add further complexity. Addressing these challenges requires advanced design methodologies, innovative operational strategies, and intelligent automation to evolve current power systems into truly smart grids.

Smart systems are at the core of this transformation, integrating renewable energy sources, digital controls, and advanced communication networks to create more efficient and resilient power grids. Research and Technology Infrastructure (RTI) are essential in this context as they provide the experimental platforms necessary for testing, validating, and refining these smart systems. By integrating diverse elements—including electrical, communication and thermal networks, market dynamics, and regulatory considerations—RTIs create cohesive environments that simulate real-world conditions. This allows researchers and engineers to evaluate performance using standardised reference scenarios, interoperability benchmarks, and harmonised cyber-physical testing procedures, ultimately ensuring that smart energy solutions are robust, efficient, and secure before field deployment.

In this context, ERIGrid 2.0 plays a crucial role. It extends and enhances the capabilities of existing RTIs by integrating physical laboratories, simulation tools, and HIL systems throughout the development of novel tools and methodologies. This integration simplifies access to multi-domain experimental environments and standardises validation procedures across various testing facilities. By addressing critical gaps in current RTIs, ERIGrid 2.0 fosters collaborative innovation and supports the development of cutting-edge solutions aligned with Europe's strategic energy targets for 2030 and 2050.

A. Acosta
RWTH Aachen University, Aachen, Germany
e-mail: andres.acosta@eonerc.rwth-aachen.de

O. Gehrke
Technical University of Denmark, Kgs. Lyngby, Denmark
e-mail: olge@dtu.dk

T. Zerihun · S. Sanchez-Acevedo
SINTEF Energy Research, Trondheim, Norway
e-mail: tesfaye.zerihun@sintef.no

S. Sanchez-Acevedo
e-mail: santiago.sanchez@sintef.no

G. Arnold
Fraunhofer Institute for Energy Economics and Energy System Technology IEE, Kassel, Germany
e-mail: gunter.arnold@iee.fraunhofer.de

J. E. Rodriguez-Seco
TECNALIA Research & Innovation, Derio, Spain
e-mail: jemilio.rodriguez@tecnalia.com

This chapter explores European energy targets and future scenarios, outlining the challenges and research objectives driving the energy transition. It specifically focuses on experiences with smart system integration and validation, drawing on feedback from Transnational Access (TA)—also called Lab Access—users and Key Performance Indicator (KPI) questionnaires. This feedback guided the development of the project tools (e.g., uAPI, configuration management tools, etc., see Chap. 6) and relative demonstrations. By learning from these experiences, RTI approaches have been refined, highlighting the advantages of integrated smart systems and documenting these benefits in forthcoming ERIGrid 2.0 reports [2]. Additionally, the chapter provides an overview of the demonstrations, showcasing the extended RTIs' capacities to deliver innovative new services.

2 User Needs and Experiences

Understanding user needs and experiences is essential for assessing the effectiveness of RTIs and ensuring that smart system integration and validation methodologies address real-world challenges. This section examines the feedback collected from TA users (see Sect. 2.1), highlighting how their insights have shaped the development of tools, methodologies, and experimental setups within ERIGrid 2.0. By systematically analysing responses from KPI questionnaires and laboratory surveys (see Sect. 2.2), it has been possible to identify key areas where improvements have been made and outline the impact of hosted projects on advancing smart energy systems. The findings presented here provide a foundation for evaluating the benefits of extended RTIs (see Sect. 3), refining future research directions, and aligning technological developments with user-driven requirements.

2.1 Experiences from Hosted User Projects

TA user projects hosted under ERIGrid 2.0 have been instrumental in testing and validating innovative technologies and methodologies for smart systems, including smart grids and electric power systems. These projects span a wide array of scientific domains, including power systems, smart grids, microgrids, energy management, electric vehicles, real-time simulation, control strategies, distributed energy resources, renewable energy integration, cybersecurity, anomaly detection, PHIL, and photovoltaic systems. A complete list is available on ERIGrid 2.0's Zenodo Community.[4] By addressing real-world challenges, these projects have not only informed the design of advanced methods and services within the extended RTIs but also underscored the critical importance of aligning technological developments with user needs.

---

[4] http://zenodo.org/communities/erigrid2.

Table 1 Overview of information collected for each user project

ID	Information requested per project
1	TA user project (UP) reference number
2	TA UP acronym
3	Brief technical summary of the project
4	User classification: Academy, Research Institute, Industry, Public Sector
5	Was this UP groundbreaking or significantly impactful for the host RTI? (Yes/No)
6	If yes, describe the new impact, development, or solution the UP helped achieve in your RTI
7	Is this an exemplary UP? (Yes/No)
8	If yes, describe what your RTI learned from the UP
9	What has the UP learned that could not have been achieved without the TA provisions?
10	What are the main outcomes or recommendations from hosting this UP in your RTI?
11	Which aspects of the TA provision process (before, during, and after UP access) worked particularly well? (Technical/Non-technical)
12	Which aspects of the TA provision process (before, during, and after UP access) did not work well? (Technical/Non-technical)

An internal survey was conducted among the RTI partners to evaluate the impact of the user projects hosted during ERIGrid 2.0. The survey collected detailed information for each project—including project reference number, acronym, a brief technical summary, user classification (Academy, Research Institute, Industry, or Public Sector), and assessments of whether the project was groundbreaking or exemplary. Table 1 summarises the information requested per project. This survey was designed to capture both technical and non-technical insights, assess the success of the TA process, and identify key outcomes and recommendations arising from the projects.

The analysis of the survey results reveals that several groundbreaking projects have made significant contributions to CPES innovation. For example, some projects have deepened the understanding and implementation of IoT-related concepts [2], while others have developed and tested new tools for upgrading microgrid SCADA systems using IoT elements [10]. Additionally, user projects have advanced decentralised energy technologies by investigating adaptive protection schemes and cyberattack scenarios in microgrids [1]. Other notable achievements include the development of dynamic models for active distribution networks based on experimental measurements, a PHIL setup for testing grid-forming inverters with voltage-type amplifiers [7], the creation of a co-simulation framework enabling real-time data sharing between geographically distributed laboratories [6], and the expansion of blockchain applications for secure energy communications [15]. A clear tendency is the application of Artificial Intelligence (AI) and machine learning techniques to smart energy systems, ranging from cybersecurity assessments [3] to monitoring the health condition of substation instrument transformers. Furthermore, other projects have evaluated the impact of cyberattacks on control systems, developed effective detection mechanisms, tested remote control strategies for PV emulation setups [11], and analysed the impact of the RES integration in power distribution networks. Finally, the laboratory access activities entailed the accelerated validation of concrete tools such as voltage state estimators, net-load forecasting tools and grid condition prognostic platforms applicable to grids and microgrids within real-time simulation environments.

These findings highlight how user feedback and rigorous testing have driven methodological and technological advancements within ERIGrid 2.0. The insights gained from these projects not only validate the extended RTI approach but also provide a strong foundation for future research and development in smart grid integration and validation.

2.2 Laboratory Questionnaires

Within the ERIGrid 2.0 project, laboratory questionnaires have been developed as an essential evaluation tool to capture detailed feedback from researchers involved in various demonstration activities (see Sect. 3). As outlined in Chap. 2, these questionnaires were meticulously designed to systematically assess the performance and impact of the extended RTIs deployed throughout the project. They are structured into two distinct categories: The first set comprises *general questions* that evaluate management and operational aspects—key concerns for non-technical stakeholders, industrial clients, and academics—while the second addresses the *specific technical requirements* of each demonstration.

The primary purpose of these questionnaires is to gather comprehensive insights into KPIs, such as system stability, accuracy, scalability, and service enhancements. This structured feedback not only validates the experimental setups and the effectiveness of the integrated tools and methodologies but also drives iterative improvements and standardisation efforts across multiple domains. In essence, the questionnaires provide a critical feedback loop that supports the continuous refinement of the smart system integration and validation processes, ensuring that the extended RTIs meet both the immediate needs of the research community and the long-term European energy targets.

For example, the PHIL demonstration (see Sect. 3.1), addresses the challenge of extending the range of stable PHIL simulations. This need is driven by the limitations of existing interface algorithms that often suffer from instability at high impedance ratios or compromised accuracy-respondents. Thus, significant improvements in both system stability and measurement accuracy were reported. These enhancements translate into better performance and reliability in simulated environments. Overall, the assessments provide a comprehensive understanding of the flexibility, operational limits, and performance of the participating RTIs, confirming that an expanded stability region and reduced error margins are crucial for validating new interface algorithms.

Similarly, the questionnaires revealed that the innovative DDPG+LSTM algorithm (see Sect. 3.2) for time delay compensation in the GDRTS setup delivered noteworthy improvements in managing variable delays. This demonstration was designed to overcome the limitations of conventional time delay compensation methods, particularly in adapting to variable delays. The assessments provide a comprehensive overview of the feasibility, applicability, limitations, and benefits of both the conventional and the DDPG+LSTM-aided approaches within the participating RTIs. Laboratory researchers observed enhanced simulation accuracy and improved power synchronisation—vital for high operational efficiency in geographically distributed systems. Despite various challenges during the demonstration, the results indicate clear progress towards the project's objectives, with significant implications for advancing research, development, and the wider adoption of GDRTS techniques.

In the Accelerated Time-to-Experiment demonstration (see Sect. 3.3), the implementation of automated configuration tools—such as VILLASnode and the CM tool (detailed in Chap. 6)—received very positive feedback. The primary goal was to streamline the setup of multi-RTI scenarios involving CHIL and PHIL. A key challenge identified was the configuration of simulators, devices, and software modules (including lab coupling tools), a process that is not only tedious but also prone to human errors, which can cause delays or even damage equipment, leading to costly repairs or replacements. Moreover, the benefits of code reuse, enabled by a harmonisation layer in experiment communications, and reductions in administrative tasks (such as the implementation of firewall rules for inter-laboratory communication) further reduce the overall time-to-experiment. Participants noted that these tools substantially reduced setup times and minimised errors, thereby lowering operational costs and streamlining experimental processes. The advantages of code reuse and enhanced communication protocols were particularly emphasised, reflecting strong support for the methodologies employed.

Overall, the questionnaires provided a rich dataset of quantitative and qualitative feedback, enabling a comprehensive evaluation of the extended RTIs. The data confirms improvements in scalability, cost efficiency, and service extensions across multiple demonstrations while also guiding future refinements. These insights ensure that ERIGrid 2.0 tools and methodologies continue to meet the evolving needs of the research community and support broader European energy targets. The complete analysis is available in [12], and a summary of key quantitative metrics extracted from the KPI questionnaires is presented in Table 2.

3 Illustrative Demonstration Activities

To validate the methodologies and tools developed within ERIGrid 2.0 (see Chaps. 2–6), a series of illustrative demonstration activities were conducted across multiple RTIs. These demonstrations were designed to address critical challenges in smart grid validation, focusing on stability, interoperability, and automation in distributed energy systems.

The first demonstration investigated the VSI method to enhance stability and accuracy in PHIL simulations, overcoming limitations in traditional interface algo-

Experiences with Smart System Integration and Validation

Table 2 Summary of quantitative metrics extracted from the KPI questionnaires

Demo	Preparation time (days)	Conduction time (days)	Improvement level (1–5)	Cost saving (1–5)	TRL
Improved Stability and Accuracy for PHIL	7	7	3	3	9
Time Delay Compensation for GDRTS	365	14	3	3	6
Accelerated Time-to-Experiment for Remote RTI Testing	168	12	3	3	~5

rithms. The second demonstration focused on analysing time delays in a multi-RTI GDRTS setup and developing methodologies for effective compensation. The third demonstration aimed to streamline the setup of distributed experiments, leveraging automated configuration tools and real-time communication frameworks to improve scalability and efficiency.

These activities were not only technical evaluations but also an essential part of user-driven research through the TA programme. The experiences of TA users provided valuable feedback on the experimental methodologies, guiding refinements in both tools and validation frameworks. By incorporating real-world user perspectives, the demonstrations ensured that the solutions developed were practical, adaptable, and aligned with the evolving needs of the research and industry community.

3.1 Extended Range High-Fidelity PHIL Testing

As power systems become increasingly complex, integrating renewable energy sources, advanced control strategies, and novel power electronics technologies, the need for rigorous testing methodologies has never been greater. PHIL simulations have emerged as a critical tool for validating these modern energy systems under realistic conditions. However, PHIL setups are prone to stability and accuracy issues, primarily due to interface dynamics and delays introduced by power amplifiers. These limitations restrict the applicability of PHIL in assessing the real-world performance of CPES technologies.

This demonstration focuses on VSI, an advanced interface algorithm designed to improve the stability and accuracy of PHIL simulations. By effectively manipulating interface signals, VSI mitigates instability while preserving the fidelity of system responses, addressing key challenges in smart system integration and validation, as discussed in the broader scope of this chapter. The demonstration directly supports the ERIGrid 2.0 objectives by developing methodologies that enhance RTI capabilities, making PHIL testing more reliable and widely applicable in power and energy systems validation.

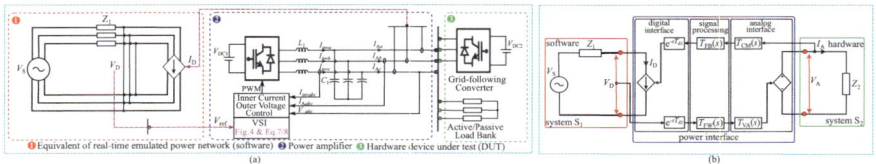

Fig. 1 **a** Schematic representation of the PHIL setup, and **b** equivalent model of the PHIL interface [9]

The experimental validation of VSI was conducted at two leading laboratories: The Power Networks Demonstration Centre and the Dynamic Power Systems Laboratory, both at the University of Strathclyde. These facilities provided the required infrastructure, including DRTS, bi-directional power amplifiers, and configurable passive load banks. The test system involved a real-time simulated power network interfaced with physical power devices through PHIL configurations. The objective was to assess how VSI enhances PHIL stability and accuracy compared to conventional methods like FBF. A schematic overview of the demonstration is presented in Fig. 1, with additional details available in [9].

The TA programme under ERIGrid 2.0 played a crucial role in shaping the development and validation of the VSI methodology. Through collaborative research projects hosted at both laboratories, visiting researchers and industry stakeholders had the opportunity to evaluate PHIL testing methodologies under real-world conditions, highlighting key challenges and areas for improvement.

Several TA users, particularly those working on renewable integration, power electronics, and microgrid control, reported frequent stability issues in PHIL experiments due to interface dynamics, latency, and amplifier-induced instabilities. Their feedback emphasized the trade-off between stability and accuracy in conventional PHIL approaches, particularly when employing LPF in FBF methods. This insight was instrumental in guiding the development of the VSI algorithm, as it directly addressed these concerns by improving stability margins while preserving system accuracy.

3.2 Time Delay Compensation for GDRTS

As modern power systems become increasingly complex and distributed, the need for accurate, real-time validation methods grows significantly. GDRTS has emerged as a powerful technique to enable collaborative experimentation across multiple RTIs. However, one of the critical challenges in GDRTS is the presence of time delays in signal exchange, which can significantly impact system stability and accuracy. This demonstration aimed to characterise and compensate for these time delays, ensuring reliable performance in geographically distributed validation environments.

The demonstration was designed to analyse time delays within a multi-RTI GDRTS setup and to develop methodologies for their compensation. The experimental setup involved multiple RTIs, where digital real-time simulators were interconnected over the internet through dedicated communication interfaces. The study focused on characterising the variability of time delays, assessing their impact on simulation accuracy, and implementing machine learning-aided compensation techniques.

The experiment employs a GDRTS framework, involving two research laboratories located approximately 3500 km apart: Dynamic Power Systems Laboratory, University of Strathclyde, and Electric Energy System Laboratory, National Technical University of Athens, as shown in Fig. 2.

To achieve these objectives, the experiment was divided into three key phases. *Time delay characterisation* involved a detailed assessment of time delays using statistical methods to evaluate their variability and uncertainty. Data was collected from multiple RTIs with measurements performed using Raspberry Pi devices deployed at different locations to monitor round-trip latency. *Accuracy analysis* and *stability assessment* focused on processing the collected time delay data to determine its distribution and impact on GDRTS fidelity. Probabilistic models quantified the likelihood of delay variations, revealing nonuniform and multimodal distributions, highlighting the need for advanced compensation strategies to mitigate their effects. Time delay compensation techniques were then developed, leveraging a machine learning-aided compensation scheme to adjust for variable yet deterministic delays in GDRTS setups. Predictive control algorithms were implemented to synchronise power signal exchanges across geographically distributed laboratories, ensuring accurate and stable real-time interactions.

User needs and experiences played a crucial role in shaping the experiment's design and execution. Feedback from previous TA projects highlighted the challenges posed by variable time delays in GDRTS. These insights underscored the

Fig. 2 Time delay measurements between geographically distributed RTIs: **a** DPSL-PNDC, **b** DPSL-ICCS [12]

necessity of developing an adaptive approach that could ensure stability and accuracy despite network-induced delays. The KPI questionnaires provided a structured evaluation of existing methods, revealing that conventional delay compensation techniques struggled to handle non-uniform and multimodal delay distributions effectively. This motivated the adoption of machine learning-based predictive control, allowing the system to anticipate and adjust for time delays dynamically.

3.3 Accelerated Time-to-Experiment for Remote RTI Testing

As distribution networks evolve into smart systems, they increasingly integrate power electronic devices, ICT infrastructure, and real-time control strategies to enhance efficiency and stability. However, this growing complexity presents new challenges for grid operators, device manufacturers, and regulatory bodies, particularly in coordinating control across DERs. Traditional testing environments struggle to capture the dynamic interactions between these elements fully, necessitating more advanced experimental approaches.

To address these challenges, this demonstration focuses on a GD-PHIL setup designed to evaluate voltage control strategies in a multi-RTI environment. The objective is to demonstrate how a CVC algorithm can manage voltage regulation devices across multiple RTIs while maintaining stability and efficiency. This aligns with the broader ERIGrid 2.0 objectives, particularly in developing tools that improve interoperability, automation, and the scalability of distributed testing environments. Concretely, voltage control is enabled directly via an OLTC or indirectly through active and reactive power injection using the PV and BESS resources.

The CVC algorithm is formulated as a constrained optimization problem. A weighted objective function minimizes the bus voltage deviation from the nominal value and the energy losses from transmission lines and penalizes the use of the OLTC to encourage the support from renewables. The problem is subject to equality and inequality constraints accounting for the energy balance, the grid elements' models, and the operational constraints. A detailed description of the CVC algorithm can be found in [8].

The GD-PHIL experiment intends to demonstrate a complex RTI integration scenario in which several participants contribute with a specific component of the grid. In addition, the main focus is on assessing the viability of the tools described in Chap. 6 to support and simplify the interoperability and automated configuration of distributed experiments.

As part of the ERIGrid 2.0 TA programme, visiting researchers participated in multi-RTI demonstrations, providing feedback on interoperability, experiment setup complexity, and system performance. The insights from these user experiences were critical in refining the automation and interoperability tools, ensuring their effectiveness in real-world applications.

4 Integration and System Validation

The integration and validation of smart energy systems require a rigorous testing approach that ensures the effectiveness, interoperability, and scalability of emerging technologies. Within ERIGrid 2.0, the conducted demonstration activities served as a foundation for evaluating key advancements in HIL methodologies, distributed control strategies, and automated experimental setups. This section presents the outcomes of these demonstrations, assessing their impact on grid stability, experimental efficiency, and the broader applicability of RTIs. Each subsection presents a comprehensive overview of the obtained outcomes, while further discussion can be found in related project results and relevant referenced papers.

4.1 Extended Range High-Fidelity PHIL Testing

A key objective of this study was to assess how the VSI methodology enhances the robustness and precision of PHIL setups while ensuring that experimental conditions closely replicate real-world power system behaviour. Beyond improving accuracy, this demonstration played a pivotal role in the integration of advanced control techniques into smart grid validation frameworks, aligning with ERIGrid 2.0's broader goal of developing standardised methodologies for RTIs to support next-generation grid technologies.

The Power Networks Demonstration Centre laboratory setup featured a DRTS from RTDS Technologies, a 270 kVA bi-directional power amplifier, and a 50 kVA passive load bank, providing a controlled yet flexible environment for validating the VSI approach. The passive load supported impedance variations between 3Ω and 100Ω, allowing for dynamic assessments of the stability margins. For the experimental analysis, an impedance ratio of $Z_1 = Z_2 = 12\Omega$ was selected to evaluate PHIL performance under realistic network conditions.

The second phase of the experiment at the Dynamic Power Systems Laboratory followed a similar setup, with modifications including a Triphase 90 kVA power amplifier and a 256-step passive load bank, extending the applicability of the VSI approach across different hardware configurations. The ability to integrate VSI within these distinct experimental environments underscored its versatility and scalability for broader research applications.

To rigorously test VSI's effectiveness, multiple scenarios were designed, comparing its performance against traditional FBF-based stabilisation methods. The study specifically focused on (i) voltage and current stability under steady-state conditions, (ii) transient response to sudden voltage fluctuations, and (iii) active and reactive power accuracy in PHIL simulations.

The experimental findings demonstrated a clear advantage of the VSI method over conventional techniques. In terms of stability improvement, while FBF methods required strict LPF with a 300 Hz cut-off to maintain marginal stability, the

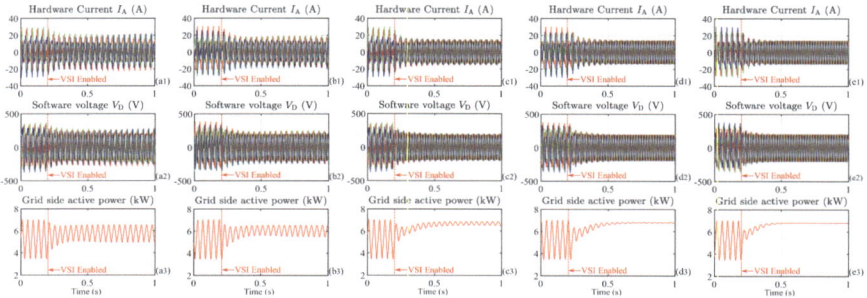

Fig. 3 PHIL experimental results comparing the FBF and VSI methods [9]

VSI approach enabled stable operation with an impedance shift of only 16%, offering a significantly wider stability region. Figure 3 presents the active and reactive power transfer with VSI. Regarding transient performance, VSI successfully maintained stable operation during an 8% voltage step change, validating its ability to enhance PHIL system resilience under dynamic conditions. In terms of accuracy enhancements, VSI significantly reduced active power oscillations and consistently exhibited greater precision compared to FBF, even when higher impedance shifting ratios (8.3% to 25%) were introduced.

This demonstration was not only a validation of VSI as a stability enhancement tool but also a critical step in integrating novel PHIL techniques into smart system RTIs. By demonstrating the applicability of VSI across different laboratory environments, this study confirmed that advanced interface algorithms can be seamlessly integrated into existing RTIs, reducing experimental uncertainties and broadening the scope of PHIL testing methodologies.

Furthermore, the results underscored the importance of structured validation frameworks for experimental research. The findings were evaluated against KPIs (see Sect. 2.2), ensuring that the method met quantifiable stability, accuracy, and scalability criteria. Additionally, TA feedback from researchers working with PHIL systems reinforced the relevance of developing standardised, interoperable validation methodologies for smart grid applications.

4.2 Time Delay Compensation for GDRTS

This demonstration was designed to evaluate the feasibility and effectiveness of a DRL-based time delay compensation method in GDRTS. Figure 4 illustrates the GDRTS delay between the Dynamic Power Systems Laboratory and Power Networks Demonstration Centre, based on 100,000 delay samples. By addressing the challenges posed by variable communication delays, the study aimed to validate the ability of an LSTM-DDPG-based approach to enhance power signal synchronisa-

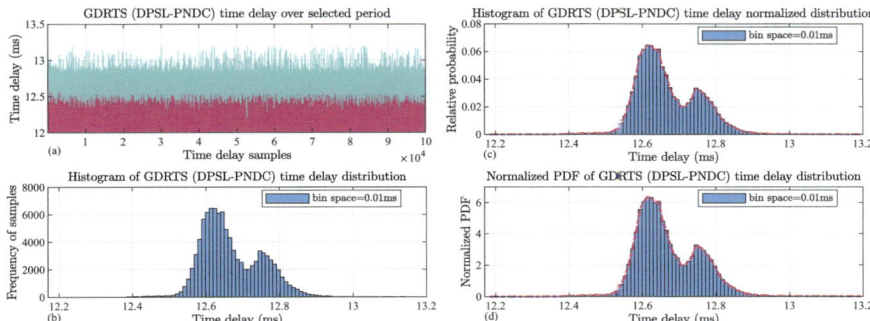

Fig. 4 Example implementation of asynchronous GDS showcasing voltage regulation in a distribution network [13]

tion and improve the stability of distributed energy systems. The experiment not only confirmed the potential of DRL for adaptive compensation but also highlighted the importance of standardised methodologies for integrating machine learning-based control strategies into multi-laboratory experimental frameworks.

The study was structured into multiple phases to systematically assess the performance of the DRL-aided delay compensation. The time delay characterisation phase involved collecting and analysing time delay data from geographically distributed RTIs, including the Dynamic Power Systems Laboratory and Power Networks Demonstration Centre, using Raspberry Pi-based latency monitoring. The accuracy analysis and stability assessment phase evaluated the impact of time delay variations on the fidelity of GDRTS by employing probabilistic models to quantify uncertainty and nonuniform distribution patterns. Finally, the time delay compensation evaluation phase tested the effectiveness of different DRL-based compensation strategies, including DDPG and an enhanced LSTM-DDPG model, in synchronising power signals across distributed laboratories.

To ensure robust validation, the experiment employed a voltage divider circuit to create a controlled testing environment for power tracking performance. The DRL agent was initially trained for fixed delays using historical delay data and then tested under both fixed and variable delay conditions. The results demonstrated that while the standard DDPG-based agent effectively compensated for static delays, it struggled with variable delays. In contrast, the LSTM-DDPG agent exhibited greater adaptability, learning to adjust dynamically to changing delays.

Key findings from the experiment include improvements in power tracking accuracy, with the LSTM-DDPG method significantly reducing active and reactive power errors compared to conventional fixed-delay compensation approaches. Phase alignment between voltage and current signals was also improved, ensuring better synchronisation of distributed power networks. Furthermore, the robustness of DRL-based compensation was demonstrated, confirming the feasibility of machine learning-based delay mitigation for geographically distributed real-time simulations. Further details can be found in related project reports [12] and Chap. 6. For the sake of brevity, plots and tabulated data are not included in this section.

These results underscore the potential of integrating artificial intelligence into real-time power system validation. By leveraging advanced reinforcement learning techniques, this study demonstrated the effectiveness of intelligent compensation methods for GDRTS, paving the way for further adoption of machine learning in distributed energy system validation.

4.3 Accelerated Time-to-Experiment for Remote RTI Testing

The integration of multiple geographically distributed RTIs was evaluated within the use case described in Sect. 3.3. A schematic representation of the laboratory setup is provided in Fig. 5. Several experiments were conducted across different RTIs in Europe, including RWTH Aachen, TECNALIA, DTU, UoS, SINTEF, and ICSS. Some of these experiments involved up to three RTIs simultaneously, utilizing various methodologies such as SIL and GD-PHIL.

A modified version of the CIGRE-MV benchmark [5] was implemented in the RTlab at RWTH. The CVC algorithm was executed in MATLAB with VILLASnode facilitating internal communication using UDP, while a Python script enabled connectivity to the JaNDER transport via the uAPI. The other participating RTIs connected to RTlab in a star topology, with those using VILLASnode employing WebRTC as the transport protocol. WebRTC streamlined connection establishment and mitigated challenges posed by restrictive firewall security policies, eliminating the need for additional tools such as VPNs and significantly reducing the setup time

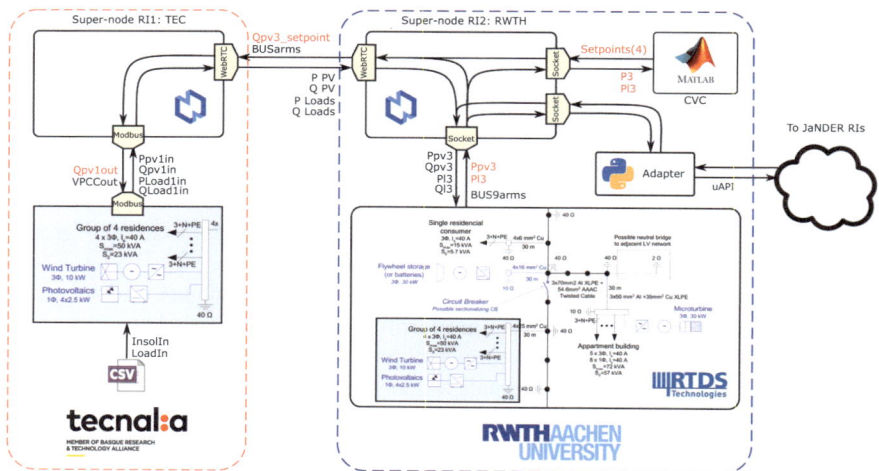

Fig. 5 Schematic representation of the GD-RTS setup between two RTIs [12]

for the distributed experiment. The only requirement for participants was the prior exchange of a session ID, which was included in the VILLASnode configuration file.

As depicted in Fig. 5, several buses of the benchmark network were designated as PCC. The corresponding RMS bus voltage was transmitted to participating RTIs following an asynchronous coupling scheme [14], under the assumption that the system frequency remained unaffected. Additionally, the voltage control setpoints computed by the CVC were transmitted, specifically the reactive power for buses with PV generation and the active/reactive power for the BESS. The participating RTIs reproduced the PCC bus voltage within their laboratories. For buses with loads, a predefined load profile was replicated. The control loop was completed by transmitting the active and reactive power values of PV and loads to RTlab, where they were routed to the RTDS. Finally, measurement signals related to active powers, PV, loads, and bus voltages were sent from the RTDS to MATLAB. As the number of participating RTIs increased beyond two, the complexity of planning, configuring, and coordinating the experiments increased significantly. To address these challenges, the CM tool (see Sect. 3.3) was deployed in a simplified version of the experiment and demonstrated in a live session [4]. The implementation of the CM tool improved the experimental setup by reducing the number of lines of code required to prepare the configuration, minimizing human errors, and streamlining verification processes.

The insights gained from this demonstration reinforce the importance of cross-laboratory collaboration in smart grid validation and demonstrate how advanced configuration management tools can enhance the feasibility and efficiency of multi-RTI experiments. These results not only validate the CVC approach in a real-world setting but also establish a framework for future scalable, automated CPES testing environments.

5 Closing Remarks

This chapter has explored the integration and validation of smart energy systems within the ERIGrid 2.0 project, highlighting the role of distributed RTIs in advancing testing methodologies and experimental validation. Through TA experiences and KPIs, the demonstrations presented here addressed critical challenges in power system stability, interoperability, and real-time control validation.

User feedback from TA experiences played a pivotal role in shaping the experimental setups, refining test scenarios, and guiding the development of tools such as uAPI, CM, and RTI adapters. These tools facilitated seamless data exchange, improved protocol compatibility, and streamlined multi-RTI experiment configuration. The lessons learned from user engagement confirmed that integrating practical feedback into the research process enhances the relevance and applicability of experimental methodologies, ensuring that RTIs remain valuable resources for advancing smart grid innovation.

Future research should continue enhancing RTI interoperability, refining experimental frameworks, and addressing remaining challenges in scalability and automation, ensuring that RTIs remain at the forefront of CPES innovation.

References

1. Ecevit MI, Ceylan O, Ozdemir A, Biricik M, Ugurlu TA, University KH (2024) OSINT and AI-based cybersecurity resilience improvement for electrical power distribution systems (ORCA). https://doi.org/10.5281/zenodo.14765955
2. ERIGrid2.0 Consortium: Erigrid2.0 zenodo community (2024). https://zenodo.org/communities/erigrid2/about. Accessed 17 Feb 2025
3. Fellner D, Thomas S (2023). Data driven detection of malfunctioning devices in power distribution systems validation (DeMaDsVal). https://doi.org/10.5281/zenodo.14515644
4. Heussen K, Gehrke O (2024) Research infrastructure automation workshop (2024). https://doi.org/10.5281/zenodo.14699238. Version Number: 2.0
5. Kotsampopoulos P, Lagos D, Hatziargyriou N, Faruque MO, Lauss G, Nzimako O, Forsyth P, Steurer M, Ponci F, Monti A, Dinavahi V, Strunz K (2018) A benchmark system for hardware-in-the-loop testing of distributed energy resources. IEEE Power Energy Technol Syst J 5(3):94–103. https://doi.org/10.1109/JPETS.2018.2861559. Conference Name: IEEE Power and Energy Technology Systems Journal
6. Lasemi MA (2022) SESA-Lab (SES-MGES). https://doi.org/10.5281/zenodo.14763644
7. Makrides G, Lopez Lorente J, Pikolos L, Bharath-Varsh R, Reisenbauer S (2022). Adv GPT Inverter Phys Demon (AGIPDem). https://doi.org/10.5281/zenodo.14506192
8. Maniatopoulos M, Lagos D, Kotsampopoulos P, Hatziargyriou N (2017) Combined control and power hardware in-the-loop simulation for testing smart grid control algorithms. IET Gen Trans Distrib 11(12). https://doi.org/10.1049/iet-gtd.2016.1341
9. Paspatis A, Kontou A, Feng Z, Syed M, Lauss G, Burt G, Kotsampopoulos P, Hatziargyriou N (2024) Virtual shifting impedance method for extended range high-fidelity phil testing. IEEE Trans Ind Electron 71(3):2903–2913. https://doi.org/10.1109/TIE.2023.3269467
10. Primas B, Miscio J (2023). Protect Optim Efficient Microgrids (POEM). https://doi.org/10.5281/zenodo.14501410
11. Rodriguez JDS, Alonso R, Maugeri G (2023). MOnitoring Failures Oper Maintenance PV (MOFOMPV). https://doi.org/10.5281/zenodo.14501690
12. Silano G, Paludetto G, Rodio C, Lazzari R, Feng Z, Kontou A, Paspatis A, Kotsampopoulos P, Gehrke O, Zerihun T, Acosta A, Rikos E, Subramaniam Rajkumar V (2024) D-JRA4.3 demonstration of the extended research infrastructure. https://doi.org/10.5281/zenodo.12796352
13. Syed M, Hoang TT, Kontou AC, Paspatis AG, Burt GM, Tran QT, Guillo-Sansano E, Vogel S, Nguyen HT, Hatziargyriou ND (2023) Applicability of geographically distributed simulations. IEEE Trans Power Syst 38(4):3107–3122. https://doi.org/10.1109/TPWRS.2022.3197635

14. Syed M, Hoang TT, Kontou AC, Paspatis AG, Burt GM, Tran QT, Guillo-Sansano E, Vogel S, Nguyen HT, Hatziargyriou ND (2023) Applicability of geographically distributed simulations. IEEE Trans Power Syst 38(4):3107–3122. https://doi.org/10.1109/TPWRS.2022.3197635. Conference Name: IEEE Transactions on Power Systems
15. Yadav A, Kishor N, Negi R, Nehru M (2022) PERformance analysis of PV integrated distribution network with combination of difFErent Control strategies and communication neTwork (PERFECT). https://doi.org/10.5281/zenodo.14516843

Open Access This chapter is licensed under the terms of the Creative Commons Attribution 4.0 International License (http://creativecommons.org/licenses/by/4.0/), which permits use, sharing, adaptation, distribution and reproduction in any medium or format, as long as you give appropriate credit to the original author(s) and the source, provide a link to the Creative Commons license and indicate if changes were made.

The images or other third party material in this chapter are included in the chapter's Creative Commons license, unless indicated otherwise in a credit line to the material. If material is not included in the chapter's Creative Commons license and your intended use is not permitted by statutory regulation or exceeds the permitted use, you will need to obtain permission directly from the copyright holder.

Education Needs, Methods and Tools

A. Kontou, P. Kotsampopoulos, K. Heussen, L. E. Ramos Perez, T. I. Strasser, E. Rikos, G. Makrides, Z. Feng, P. Karafotis, A. Paspatis, M. Syed, G. Lauss, M. Calin, and N. Hatziargyriou

Abstract This chapter addresses the evolving educational needs in the energy sector, emphasizing interdisciplinary training and the integration of modern teaching methods and tools. It highlights the importance of open-source e-learning resources, virtual and remote laboratories, and interactive educational tools in fostering practical skills and theoretical knowledge. Additionally, it underscores the contributions of the ERIGrid 2.0 project in developing comprehensive training programs and resources to equip the next generation of energy professionals with the necessary expertise to meet industry demands.

A. Kontou (✉) · P. Kotsampopoulos · N. Hatziargyriou
National Technical University of Athens, Athens, Greece
e-mail: alkistiskont@mail.ntua.gr

P. Kotsampopoulos
e-mail: kotsa@power.ece.ntua.gr

N. Hatziargyriou
e-mail: nh@power.ece.ntua.gr

K. Heussen
DTU Danmarks Tekniske Universitet, Roskilde, Denmark
e-mail: kheu@dtu.dk

L. E. Ramos Perez
European Distributed Energy Resources Laboratories (DERlab) e.V., Kassel, Germany
e-mail: leonard.ramos@der-lab.net

T. I. Strasser · G. Lauss · M. Calin
AIT Austrian Institute of Technology, Vienna, Austria
e-mail: thomas.strasser@ait.ac.at

G. Lauss
e-mail: georg.lauss@ait.ac.at

M. Calin
e-mail: mihai.calin@ait.ac.at

E. Rikos
CRES Centre for Renewable Energy Sources and Saving, Pikermi, Greece
e-mail: vrikos@cres.gr

© The Author(s) 2025
T. I. Strasser et al. (eds.), *European Guide to Smart Energy System Testing*, SpringerBriefs in Energy, https://doi.org/10.1007/978-3-031-99451-7_9

1 Interdisciplinary Training in Modern Energy Systems

The energy landscape has undergone a rapid transformation and digitalization over the past decades, driven by emerging technologies and scientific advancements. This accelerated transition presents two key workforce challenges: (i) the sector's rapid expansion has outpaced the availability of skilled professionals, and (ii) the rise of new technologies and disciplines has created a shortage of specialists capable of sustaining growth and innovation at the required pace [7]. The shortage of qualified professionals underscores the need for continuous skills development and specialized training to ensure a workforce that can meet the sector's evolving demands [8]. Thus, a crucial aspect that needs attention is the modernization and adaptation of power systems education and training. Specifically, training and educational programs should be designed to equip both current and future engineers and researchers with a comprehensive understanding across multiple domains, including electric power, thermal energy, and most critically, ICT [5].

To effectively address skill shortages, comprehensive training and continuous skills development are essential at all stages of education and across all age groups. Until now, efforts have primarily focused on upgrading and improving education at the university level. However, it is suggested here that preparing the next generation of researchers, engineers, and professionals requires a broader approach. This includes updating and enhancing curricula across all educational levels, from primary and secondary school to undergraduate, master's, and Ph.D. programs, while also ensuring the provision of open-source tools and training materials. Providing accessible learning opportunities not only fosters inclusivity but also supports individuals looking to transition careers or expand their expertise in emerging fields.

To enrich the educational experience, foster new skill acquisition, and develop a strong interdisciplinary foundation, it is crucial to leverage a diverse range of teaching methods and learning platforms. Traditional lectures, seminars, and workshops

---

G. Makrides
University of Cyprus, Nicosia, Cyprus
e-mail: makrides.georgios@ucy.ac.cy

Z. Feng
University of Strathclyde, Glasgow, UK
e-mail: zhiwang.feng@strath.ac.uk

P. Karafotis
HEDNO Hellenic Electricity Distribution Network Operator, Athens, Greece
e-mail: p.karafotis@deddie.gr

A. Paspatis
Manchester Metropolitan University, Manchester, UK
e-mail: a.paspatis@mmu.ac.uk

M. Syed
WSP Energy Advisor, Glasgow, UK
e-mail: mazher.syed@wsp.com

should be complemented by e-learning tools and digital resources, such as webinars, online simulation tools, e-books, and educational videos, aligning also with the recommendations of [6]. Furthermore, hands-on experiences based on hardware setups and advanced laboratory validation methods like co-simulation, RTS, and HIL techniques play a vital role in bridging the gap between theory and practical application, reinforcing knowledge through experiential learning. Using the recent technological advancements and means and applying the new trends in laboratory education is fully aligned with the recommendations and best practices suggested by the prominent collaborative effort [6].

2 Open-Source E-Learning Education

Given the interdisciplinary nature of power and energy systems, learners need to engage with a diverse range of tools and concepts across multiple domains. To support this, innovative e-learning educational methods and online tools must be developed to integrate knowledge from various fields, helping learners to comprehend the interactions between different components within intelligent energy solutions. Furthermore, the creation of open-source educational resources promotes accessibility and inclusivity, ensuring that high-quality content and tools are available to anyone with an internet connection, regardless of location or financial resources.

To this end, a wide range of digital open-source platforms are available and offer valuable opportunities for creating interactive training and educational materials. Virtual and remote laboratories have revolutionized smart grid applications by providing access to emulated or remotely controlled physical laboratory infrastructures eliminating the need for on-site presence while enabling hands-on experimentation and learning. Software and simulation tools play a crucial role in engineering education, enabling learners to understand and apply methods for designing, analyzing, and optimizing smart energy systems. Beyond enhancing knowledge of smart grid operations, these tools also help to develop ICT skills, fostering interdisciplinary competence essential for modern energy systems engineers and professionals. In recent years, Massive Open Online Courses (MOOCs) have gained significant attention as comprehensive courses that incorporate interactive elements such as video lectures, quizzes, and assessments. Leading universities, industry leaders, and research initiatives offer MOOCs to deliver high-quality, free, or low-cost training in smart energy education. These courses scale learning to a global level, equipping and empowering the next generation of energy professionals with essential knowledge and skills. Furthermore, webinars and videos, can provide step-by-step guidance on using tools, developing smart energy solutions and applying technologies for energy system modeling and analysis based on case studies. At the same time, they provide learners with valuable opportunities to engage with educators, fostering a more dynamic and interactive learning experience. The following subsections introduce newly developed exemplary training opportunities based on the modern training methods outlined above.

2.1 Virtual and Remote Laboratories

The rapid evolution of modern and smart energy systems requires innovative approaches for testing and verifying research methods and tools. Virtual and remote laboratories are critical in this transformation since real-time access to experimental setups, simulations, and energy data is provided without physical presence requirements. In this context, virtual and remote laboratories leverage standardized protocols, cloud-based architectures, and APIs to enable seamless collaboration and cross-platform compatibility among researchers and the industry.

One particular aspect related to developing researchers' skills addresses the impact of environmental conditions on the physical layer of CPES. To address this aspect, a virtual lab called the Environmental Measurements Platform (EMP) is proposed [1]. Developed by CRES, the platform offers access to real-world environmental and electrical measurements. The specific emulated laboratory grants access to selected environmental and operational data from CRES-owned and operated PV systems on its premises in Pikermi, Greece. Download of historical data in *.csv form for further use and post-processing is also available.

Also developed by CRES, the Remote Microgrid Control Laboratory aims to expand the interaction of the user with the online EMP. This remote lab access complements the EMP in the sense that it enables remote control based on the operating/environmental conditions of the experimental microgrid located on the same premises. The remote lab offers controllability of three different DER units, namely one battery inverter and two small PVs. With the dedicated GUI depicted in Fig. 1, the user can monitor and control basic electrical quantities to maintain the energy balance of the microgrid.

Another exemplary CPES training is the virtual laboratory at the University of Cyprus, which is an interactive platform for monitoring and analyzing the operation of microgrids equipped with smart meters, Programmable Automation Controllers (PACs), and cloud-based supervision systems. Specifically, the virtual laboratory

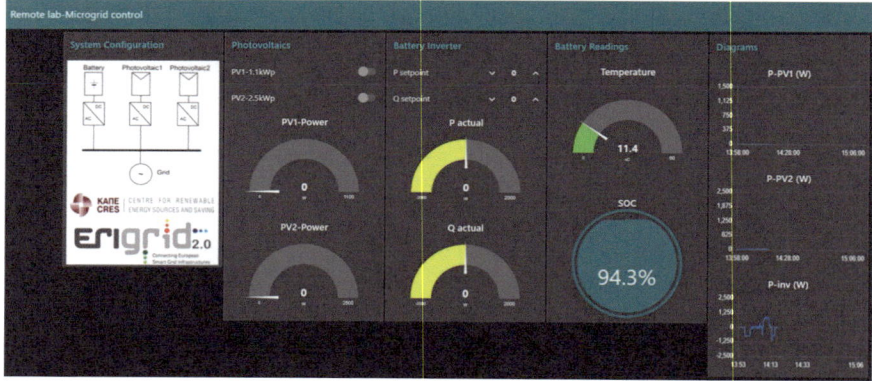

Fig. 1 Remote laboratory for microgrid energy balancing control

allows the demonstration of advanced interoperability and real-time energy management concepts of smart grids by integrating and centrally controlling DERs with the use of digitally-enhanced methods, edge computing, and cloud-based automation.

In the context of microgrids, the virtual lab grants users virtual access to Supervisory and Control System (SCADA) dashboards that stream real-time data (resolution of a second) from the installed network of smart meters within the University campus microgrid. These meters collect critical energy parameters, including voltage, frequency, and power measurements (apparent, active, and reactive), as well as power quality indicators such as Total Harmonic Distortion (THD), Crest- and K-factor. Through Modbus TCP communication interfaces, users can observe, analyze, and export historical consumption and generation data, gaining insights into real-time energy supervision and observability. In addition, the virtual lab platform presents solar-plus-storage system energy management concepts since access is provided to the supervisory dashboard of a test-bench PV and battery storage system. In particular, the setup allows users to monitor real-time data from smart PV and battery storage inverters, capturing parameters such as charging/discharging power, inverter temperature, voltage, and operational status. By leveraging Modbus TCP Vendor-specific and SunSpec protocols, the virtual lab showcases interoperability and real-time observability of DERs.

2.2 Interactive Educational Tools

Interactive educational tools play a vital role in advancing the understanding and management of modern smart energy systems by facilitating hands-on engagement within virtual environments. A central feature of these tools is the ability to model and analyze the behavior of energy systems, alongside the application of advanced control techniques for optimized operation. At the same time, learners gain exposure and develop expertise in complementary fields such as programming, information and communication technologies, artificial intelligence, fostering well-rounded, multidisciplinary skill sets.

One of the tools developed with the scope to facilitate understanding of the physical layer of CPES is the Power Flow Calculator (PFC) [3]. This tool was developed by CRES as a Python platform, which provides convenience in use combined with open-source characteristics. The tool serves the scope of steady-state analysis of power systems, thus emphasizing operational characteristics of distributed power systems that incorporate large shares of DER. As an educational tool, the PFC can be used by researchers who want to familiarize themselves with the concepts of power flow calculation. From the usage viewpoint, thanks to its open source nature, the tool can be used either as an autonomous application or incorporated in other compatible platforms. The source code was developed using Python v3.6.4 and is structured into three parts, one related to the input data, one containing the power flow calculation libraries, and the output data generation module. The user can interact with the tool by introducing/exporting data related to voltages, active and reactive power, etc. by means of I/O files in *.txt format.

Complementing the scope of the PFC a second tool related to the techno-economic analysis of microgrids and island power systems is also proposed. The specific tool is named Flexxy [2] and was also developed by CRES to expand the skills of young engineers to understand and design microgrids from an energy balance point of view. The latter scenario is becoming very important considering the developments in the energy sector related to Energy Communities and other similar organizational schemes. In terms of use, the tool is offered as an open-license application equipped with a Graphical User Interface (GUI) that enables the easy parametrization of the model. Moreover, the calculation results can be exported for post-processing.

Along this context are the interactive tools developed at the University of Cypress for the demonstration of advanced interoperability and data-driven control of modern DER management within microgrids. One of the tools developed to leverage data-driven machine learning principles for smart grid applications is the solar-specific predictive modeling and forecasting tool. The primary scope of this tool is to create digital twin predictive performance models and solar generation forecasts, enabling day-ahead and hour-ahead energy predictions. Specifically, the tool facilitates the construction of high-performing data-driven predictive models for PV systems, employing a supervised learning approach and multiple machine learning techniques such as artificial neural networks, support vector regression, and regression trees. Developed using the R scripting statistical computing programming language, the tool allows users to build, benchmark, and optimize predictive models by adjusting input features, learning regimes, and hyperparameters.

2.3 *Holistic Smart Energy System Validation MOOC*

The growing need to equip young engineers and researchers with hands-on, practical skills has elevated the importance of experimentation. Furthermore, bridging the gap between academia and industry requires advanced laboratory methods that not only capture the complex dynamics of different domains but also enable testing under near-real conditions while minimizing the risk of hardware damage. Moreover, for successful experimental validation that ensures all relevant aspects are thoroughly tested, it is essential not only to develop technical and multi-disciplinary hands-on skills but also to systematically organize test cases into well-structured and comprehensive segments. To address these needs, the ERIGrid 2.0 MOOC on "Holistic Smart Energy System Validation" is an online training program designed to provide in-depth insights into the latest advancements in smart energy system validation. It covers key methodologies such as co-simulation, HIL, GDRTS, and multi-domain experiments. Additionally, the course provides valuable training on the design of complex multi-domain and multi-research infrastructure tests, ensuring a systematic and reproducible validation process.

Education Needs, Methods and Tools

The course begins with an introduction to ERIGrid 2.0, emphasizing its role in advancing smart energy system research and particularly focusing on key trends in energy system digitalization and the importance of integrated research infrastructures. It also discusses the skill gaps in the sector and the role of e-learning in bridging them. Next, the course explores multi-energy networks and introduces co-simulation techniques for modeling complex systems. Emphasis is placed on the mosaik co-simulation framework, including a hands-on tutorial and a real-world benchmark model demonstrating the integration of electricity and heat grids. As the course progresses, learners are introduced to HIL validation methods for modern energy systems, including CHIL and PHIL. The module also covers GDRTS for integrating remote labs to enhance testing capabilities. Then, the course focuses on multi-laboratory experiments, discussing methodologies and tools for conducting experiments across different infrastructures, with real-world examples of geographically distributed simulations utilizing cloud and edge computing. Another key module covers the design and documentation of complex test setups using the HTD methodology, which ensures systematic testing and the reproducibility of results. Finally, the course presents industry insights, focusing on the practical applications of the previously presented methods. It highlights industry validation needs and success stories of lab access initiatives that demonstrate how businesses and institutions have leveraged advanced testing facilities. By the end of the course, learners gain a thorough understanding of smart energy system validation methodologies, equipping them with the skills to apply these concepts in both academic and industry settings.

3 Laboratory Education

Laboratory education in the field of electric power systems is typically conducted using software tools and dedicated hardware setups. When studying power electronic components or electric machines, hands-on exercises using physical models, sub-models, or real hardware are utilized. Since the operation of physical hardware devices is costly, time-consuming, and requires a high level of safety and know-how concerning the user, only limited or simplified hardware systems are applied for laboratory training. Offline simulation tools can be used for numerical and non-time critical investigations on generic electric power systems. When more specific requirements of simulation methods referring to real-time capability are needed, real-time HIL simulation methods can be included for educational purposes. The operation principle and the main difference between real-time and non-real-time simulation are defined and explained in [4].

For better understanding and applying real-time based simulation methods for education purposes, it is recommended to use and study [9]. Here, the vision for a first comprehensive standardization framework on HIL simulation has been realized by members of the IEEE WG P2004. This recommended practice considers HIL simulation in which a closed loop between software models and hardware of interest is formed and simulations are executed in real-time. This Standard includes a description of the key characteristics of real-time models and simulation along with recommendations on how to apply them in an HIL simulation application. Best practices and recommendations on how to establish and execute a successful HIL simulation are explained. Many example applications of CHIL and PHIL simulations, as well as a deep dive on important topics such as circuit decoupling techniques, accuracy, stability, HIL interfaces, power amplifiers, and power HIL protection techniques, are provided. Novel laboratory modules for education in smart energy systems, based on complete hardware setups and real-time simulation techniques, are presented below.

3.1 Example 1: Hierarchical Control of Microgrids

This laboratory module, developed by the Institute of Communication and Computer Systems, introduces students to the hierarchical control of inverter-dominated microgrids, with a focus on the primary and secondary control layers. As distributed generation technologies, such as microturbines, PVs, fuel cells, and small wind turbines, become increasingly integrated, traditional distribution networks are evolving from passive consumers into active prosumers, enabling bidirectional power flow. To ensure the effective integration of these microgeneration sources and variable loads, local control systems are essential for managing and coordinating assets efficiently. Microgrids exemplify this concept, offering modularity and flexibility to operate both in grid-connected and islanded modes, thereby enhancing the robustness, resilience, and efficiency of smart energy systems. The module offers hands-on experience utilizing RTS environment and introduces the learners to HIL techniques. These methods equip students with applied knowledge and a deeper understanding of component and system-level operations, essential for both academic and industrial contexts. At the end of the module, students have gained a solid understanding of primary control strategies, such as conventional and generalized droop control, as well as centralized and distributed secondary control schemes. Furthermore, the module offers practical experience with real-time simulation and fosters an intuitive understanding of design trade-offs, a critical aspect of engineering practice.

3.2 Example 2: GDRTS to Educate Young Engineers

As a core approach supporting high-fidelity validation of complex power systems, the GDRTS method has been leveraged to support the education and training of students (cf. Fig. 2), the majority of whom have worked as electrical engineers in the UK power industry. At the University of Strathclyde, this technique has been employed to support students' final-year group projects, guiding students through the entire process, from fundamental power system modeling to advanced power electronics

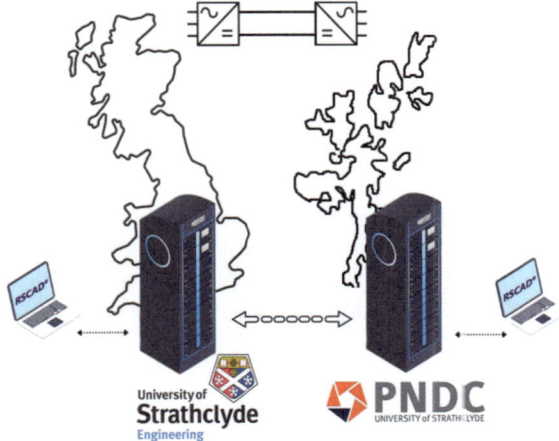

Fig. 2 Applying GDRTS approach for laboratory education

system modeling, progressing through offline and real-time simulations to real-time co-simulation across geographically distributed laboratories.

This project, titled "Real-Time Detailed Great Britain Power System Model Enabled by Geographically Distributed Laboratories," builds upon the aggregated 9-bus UK power network model to conduct RTS-based steady state and transient emulation of selected UK network events. This was subsequently followed by the design and development of hydrogen and fuel cell energy system models, which were integrated within the 9-bus UK power network model to simulate and analyse, in real-time, for the years 2024, 2035, and 2043 evolving technological landscape and system transformation under future energy scenarios in the North of Scotland. In addition, a wind farm model of the Shetland network was developed and integrated into the refined UK network model through a multi-terminal HVDC (MTHVDC) model. The wind farm model, along with MTHVDC models, were hosted in an RTS-based system at one laboratory, while the refined UK network model was hosted in an RTS-based system at another laboratory. Both systems, geographically separated by approximately 14 miles, were interconnected using the GDRTS method, enabling a comprehensive and high-fidelity representation of the evolving UK power network.

Through this project, students not only expand their knowledge in electric power engineering but also gain hands-on experience in power system design, model development, simulation, and renewable energy-based power system analysis. Furthermore, the project provides students with the opportunity for on-site visits to real-world power networks, offering valuable insights into the existing UK power infrastructure and Scotland's future energy decarbonization landscape. These experiences play a crucial role in shaping their careers, preparing them to become the next generation of professionals in the power industry.

4 Trainings and Workshops

Organizing training events, such as workshops and training schools, at renowned universities and research institutes equipped with advanced research infrastructure offers young researchers and professionals a valuable opportunity to engage in hands-on training and gain access to cutting-edge laboratory facilities. Workshops are typically shorter in duration, while training schools extend over multiple days, yet both emphasize active participation in a structured and guided setting. Their impact is maximized when they focus on the practical application of concepts, integrating live demonstrations and laboratory sessions that utilize specialized tools, equipment, and software to enhance learning and skill development. As part of the events, visits to innovative real-world pilot sites and living labs were highly encouraged, providing a unique opportunity for participants to bridge theoretical knowledge with practical application in real operational environments. To further motivate and enrich the learning experience, these events incorporated industry sessions and visits to industrial facilities, exposing participants to real-world applications and challenges. Such direct engagement with experts provides a dual benefit:

- Enhanced networking opportunities, fostering collaborations and mentorship among participants.
- Industry-driven insights, where organizations can communicate their skill development needs, ensuring that training programs align with industry demands and equip participants with relevant expertise.

By bridging academic knowledge with industrial practice, training events create a dynamic learning environment that not only strengthens technical skills but also prepares participants for real-world challenges and career growth.

Last but not least, creating opportunities and encouraging staff exchanges for students and researchers from different research institutes and universities serves as a powerful mechanism for enhancing collaboration, knowledge transfer, and skill development. These exchanges provide participants with the unique opportunity to share experiences, gain exposure to new methodologies, operational workflows, and problem-solving approaches, and expand their expertise through hands-on engagement in diverse research environments.

5 Educational Programme Management Methodology

As already highlighted, the power and energy sector has experienced rapid digital transformation, prompting universities, research institutes and industry stakeholders to rethink their educational and training approaches. To keep pace, educators must quickly evolve curricula and teaching methods to prepare the next generation of engineers. Designing new educational programmes is already complex, but the challenge grows when large research consortia must also deliver high-quality education and training within tight project timelines and scopes.

While traditional educational programme management is well-suited for academic institutions, applying it to research projects presents unique challenges due to fundamental differences in structure, purpose, and implementation. Unlike traditional educational institutions that operate within a hierarchical framework, research projects are typically distributed across multiple partners and lack centralized authority, making standard top-down management approaches less effective. Moreover, while educational programmes define learning outcomes in advance, research projects deal with evolving content, where competencies and training objectives emerge dynamically throughout the project. Another challenge lies in assessment-formal tools like ECTS credits are rarely applicable, and evaluating tacit, hands-on skills common in research is difficult. Additionally, there is often no continuity in the training audience, as participants in research-based education may not follow a sequential learning path, unlike students in structured curricula. Finally, training activities in research projects are usually short-term and tied to the project's lifecycle, in contrast to the long-term, ongoing nature of formal educational programmes. These differences highlight the need for a more flexible and responsive educational management approach tailored specifically to the research context.

For the above reasons, a methodology for educational and training activity coordination that combines both top-down and bottom-up approaches is proposed to accommodate the dynamic and distributed nature of research projects. Key elements of programme management are adopted, but unlike traditional hierarchical models, this agile framework allows for the emergent definition of competencies and learning outcomes through collaborative input. The process (as shown in Fig. 3) begins with identifying the project's technical scope, priorities, and target audience, involving all

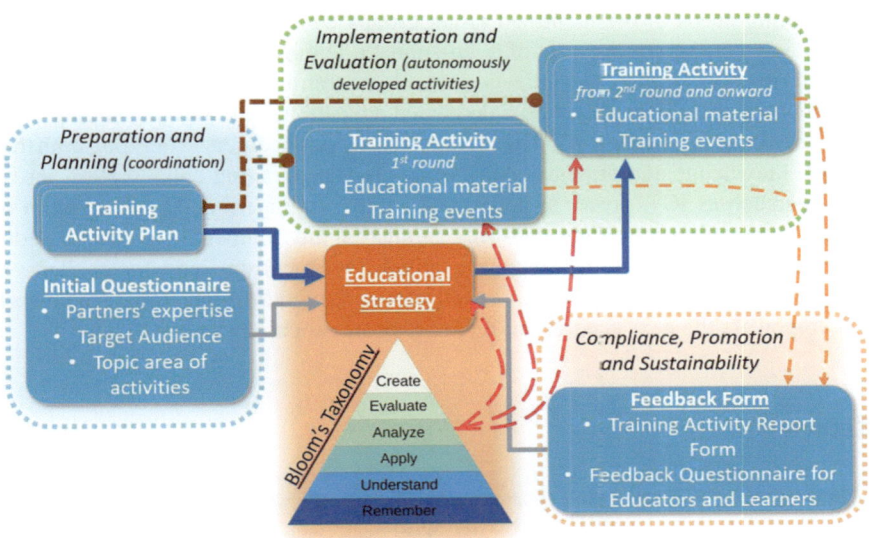

Fig. 3 Overview of the educational programme management methodology [5]

partners in defining key training topics. Next, training activities are developed and documented using standardized templates to ensure clarity and traceability. A cyclical reporting and feedback process then captures delivery details, learning objectives, and participant evaluations, which guide continuous improvement. Finally, an analytical evaluation of training impact is conducted to assess progress and adjust the strategy as needed. This iterative process ensures the educational strategy remains aligned with the project's evolving goals and supports the effective transformation of research outputs into meaningful training outcomes.

6 Conclusions

Within ERIGrid 2.0, the educational and training needs of the identified target groups were addressed through a holistic and structured approach. A broad suite of resources was developed, including open-source e-learning tools, virtual and remote laboratories, instructional videos, MOOCs, and real-time simulation environments. In addition, hands-on training modules, workshops, and training schools were delivered, all supported by a dedicated educational and training strategy that provided a cohesive framework. These resources help learners build both theoretical understanding and practical expertise while fostering collaboration across key domains such as ICT, power systems, and control engineering. Altogether, these efforts contribute to a comprehensive, modern educational ecosystem that equips the next generation of energy professionals with the skills and knowledge needed to meet evolving industry demands.

This chapter underlines the growing need for accessible, interdisciplinary education in the field of smart energy systems. Through ERIGrid 2.0, the educational and training needs of the identified target groups were approached holistically, by developing a comprehensive set of open-source e-learning tools, virtual and remote laboratories, videos, etc., the delivery of hands-on training modules, workshops and training schools, all supported by a dedicated educational and training strategy that provides a cohesive framework. These resources support learners in building both theoretical knowledge and practical skills while also encouraging collaboration across domains such as ICT, power systems, and control engineering. The integration of MOOCs, modeling tools, and real-time simulation environments ensures that education remains relevant, engaging, and aligned with current industry challenges. Overall, these efforts contribute to a well-rounded, modern educational framework that empowers the next generation of energy professionals.

References

1. Environmental measurements platform (2022). http://www.cres.gr/pv_laboratory/pages/measur/PVmeas_ltm_en.html. Accessed 06 Apr 2025
2. Erigrid2/energy-analysis-tool (2022). https://github.com/ERIGrid2/energy-analysis-tool. Accessed 06 Apr 2025

3. Erigrid2/load-flow-tool (2022). https://github.com/ERIGrid2/load-flow-tool. Accessed 06 Apr 2025
4. Faruque MO, Strasser T, Lauss G, Jalili-Marandi V, Forsyth P, Dufour C, Dinavahi V, Monti A, Kotsampopoulos P, Martinez JA et al (2015) Real-time simulation technologies for power systems design, testing, and analysis. IEEE Power Energy Technol Syst J 2(2):63–73
5. Kontou A, Heussen K, Perez LR, Karafotis PA, Syed M, Paspatis A, Rikos E (2023) Educational programme management methodology for research projects. In: 2023 IEEE 2nd industrial electronics society annual on-line conference (ONCON), pp 1–6. https://doi.org/10.1109/ONCON60463.2023.10430995
6. Kotsampopoulos P, Hatziargyriou N et al (2024) Innovative teaching methods for modern power and energy systems. https://resourcecenter.ieee-pes.org/publications/technical-reports/pes_tp_tr120_peec_032624
7. Kotsampopoulos P, Hatziargyriou N, Strasser TI, Moyo C, Rohjans S, Steinbrink C, Lehnhoff S, Palensky P, Meer AA, Morales Bondy DE, Heussen K, Calin M, Khavari A, Sosnina M, Rodriguez JE, Burt GM (2017) Validating intelligent power and energy systems—a discussion of educational needs. In: Mařík V, Wahlster W, Strasser T, Kadera P (eds) Industrial applications of holonic and multi-agent systems. Springer International Publishing, Cham, pp 200–212
8. Kotsampopoulos PC, Kleftakis VA, Hatziargyriou ND (2017) Laboratory education of modern power systems using phil simulation. IEEE Trans Power Syst 32(5):3992–4001. https://doi.org/10.1109/TPWRS.2016.2633201
9. Lauss G, Lundstrom B et al (2025) (Members IEEE WG P2004): IEEE std 2004–2025: hardware-in-the-loop (hil) simulation based testing of electric power apparatus and controls (approved 12 Feb 2025). https://ieeexplore.ieee.org/servlet/opac?punumber=10806535

Open Access This chapter is licensed under the terms of the Creative Commons Attribution 4.0 International License (http://creativecommons.org/licenses/by/4.0/), which permits use, sharing, adaptation, distribution and reproduction in any medium or format, as long as you give appropriate credit to the original author(s) and the source, provide a link to the Creative Commons license and indicate if changes were made.

The images or other third party material in this chapter are included in the chapter's Creative Commons license, unless indicated otherwise in a credit line to the material. If material is not included in the chapter's Creative Commons license and your intended use is not permitted by statutory regulation or exceeds the permitted use, you will need to obtain permission directly from the copyright holder.

Standardisation, Policies and Interoperability

M. Calin, G. Lauss, L. E. Ramos Perez, and T. I. Strasser

1 Standardisation in Energy Systems

Standardization in energy systems is crucial for several reasons. Firstly, it ensures interoperability, allowing different components of an energy system to work together seamlessly, regardless of the manufacturer. This is especially important for renewable energy sources like solar panels or wind turbines that need to integrate with the existing grid. Secondly, standardized practices and equipment help maintain high levels of safety and reliability in energy systems, which is essential to prevent accidents and ensure a consistent supply of energy. Thirdly, standardization reduces the costs associated with custom solutions and helps manufacturers scale production, ultimately lowering costs for consumers. Moreover, it provides a foundation for innovation by establishing a common framework within which new technologies may be developed, accelerating the adoption of more efficient energy solutions. Standardization also aids in monitoring and reducing the environmental impact of energy systems by standardizing the measurement and reporting of energy use and emissions. Finally, for countries and companies operating within the global market, standardization ensures that their products and systems are compatible with international standards, facilitating global trade and cooperation. In essence, standardization acts as the common language that allows all parts of the energy system to communicate effectively, ensuring that everything runs smoothly and efficiently.

M. Calin (✉) · G. Lauss · T. I. Strasser
AIT Austrian Institute of Technology, Vienna, Austria
e-mail: mihai.calin@ait.ac.at

G. Lauss
e-mail: georg.lauss@ait.ac.at

L. E. Ramos Perez
European Distributed Energy Resources Laboratories (DERlab) e.V., Kassel, Germany
e-mail: leonard.ramos@der-lab.net

© The Author(s) 2025
T. I. Strasser et al. (eds.), *European Guide to Smart Energy System Testing*,
SpringerBriefs in Energy, https://doi.org/10.1007/978-3-031-99451-7_10

Standardization is essential for smart energy systems and CPES-based RTIs as it ensures interoperability among diverse components, facilitating seamless communication and collaboration. It promotes data consistency through standardized formats and protocols, enabling effective data sharing and analysis. Additionally, standardization enhances scalability, allowing systems to expand and integrate new technologies with ease while also driving cost efficiency by minimizing the need for custom solutions. It aids in regulatory compliance, fosters safety and reliability, and encourages innovation by providing a common framework for development. Furthermore, standardized systems enhance user trust and acceptance, promote knowledge sharing and collaboration among stakeholders, and contribute to sustainability by optimizing energy management practices. Ultimately, standardization underpins the effectiveness and advancement of smart energy systems.

The International Renewable Energy Agency (IRENA) discusses the role of international standardization in renewable energy, highlighting its importance for interoperability, safety, and innovation. The ISO 50001 standard provides a framework for energy management systems, emphasizing the benefits of standardization in improving energy use, safety, and cost efficiency [1].

Current standards in energy system technologies focus on enhancing efficiency, sustainability, and interoperability across various platforms. The integration of renewable energy sources, such as solar and wind, is guided by standards that ensure compatibility with existing grids and technologies. Smart energy systems technologies are increasingly emphasized, promoting real-time data exchange and management for optimized energy distribution. Additionally, cybersecurity standards are critical, given the reliance on digital infrastructures. Internationally recognized frameworks, like those from the International Electrotechnical Commission (IEC) and the Institute of Electrical and Electronics Engineers (IEEE), help shape these standards, facilitating global collaboration while addressing regional energy needs and environmental concerns.

2 Policy Frameworks Supporting Standardisation and RTIs

2.1 Relevant National and International Policies

The European Union is committed to establishing and sustaining world-class RTIs through a comprehensive framework that includes various initiatives and regulations. Central to this effort is the development of European RTIs, which encompass major scientific equipment, extensive data collections, advanced computing systems, and robust communication networks [5]. The Horizon Europe program plays a crucial role by supporting the creation of new pan-European RTIs [6], facilitating access for transnational users, and promoting the advancement of innovative scientific instrumentation and methodologies. Complementing these initiatives is the European Research Infrastructure Consortium (ERIC) [4], a legal entity designed to streamline the establishment and operation of research infrastructures [10], thereby integrating them as essential pillars within the European Research Area (ERA) [3]. Furthermore, the ESFRI [17] contributes to this landscape by creating a strategic roadmap that outlines investment priorities for the next 10 to 20 years. Together, these policy frameworks ensure that the EU remains at the forefront of innovation and research, fostering the development of cutting-edge technologies and maintaining its global competitiveness [7].

On the international stage, the Organisation for Economic Co-operation and Development (OECD) offers a framework for optimizing national research infrastructures, including models for effective management and user engagement [8, 15, 16]. Additionally, international standardization bodies such as the International Organization for Standardization (ISO) [11] and the IEC [12] are pivotal in developing global standards that ensure safety, reliability, and efficiency across various sectors, including energy. Initiatives like the Global Research Infrastructure (GRI) framework [2, 9] further promote international collaboration in the development and utilization of research infrastructures. Collectively, these policies and frameworks foster innovation and enhance the global competitiveness of research infrastructures, ensuring their effective operation and standardization at both national and international levels.

2.2 Influence of Policies on Standardisation

The influence of policies on standardization is multifaceted and significantly shapes how standards are developed, adopted, and implemented across various sectors. One key aspect is the establishment of regulatory frameworks, where governments can set mandatory standards in industries such as healthcare, construction, and telecommunications to ensure safety, quality, and interoperability. In the context of ERIGrid 2.0, this regulatory environment is crucial for promoting standards that facilitate the integration of innovative energy technologies.

In the realm of international collaboration, ERIGrid 2.0 played a pivotal role through its Lab Access initiative, which supported the harmonization of standards across countries. This initiative facilitated smoother international partnerships by effectively reducing technical barriers that often hindered cross-border operations. By engaging in mutual recognition agreements, Transnational Access simplified processes for organizations operating in multiple jurisdictions, allowing for the acknowledgment and acceptance of standards from different regions.

Furthermore, the policies that promoted research and development within the ERIGrid 2.0 framework, particularly through Transnational Access, paved the way for the creation of new standards. This was especially crucial in rapidly evolving fields such as renewable energy technology, where innovation was paramount. The initiative also fostered the adoption of emerging technologies, including artificial intelligence in grid management, ensuring that these advancements were aligned with international standards and practices. Through its Lab Access component,

ERIGrid 2.0 not only enhanced interoperability but also strengthened the global energy landscape by promoting collaborative development and standardization across borders.

Collaboration and partnerships were vital for effective standardization; policies that promoted engagement among stakeholders, including government, industry, and academia, led to more inclusive and widely accepted standards. The ERIGrid 2.0 project emphasized this collaborative approach by facilitating partnerships that addressed both public interests and private sector capabilities. On a global scale, ERIGrid 2.0 influenced standardization by participating in international standard-setting organizations, contributing to the shaping of standards adopted worldwide. International agreements and treaties related to energy and environmental policies also dictated national policies concerning standardization, and ERIGrid 2.0 played a critical role in ensuring that these standards were met by advancing innovative energy solutions that complied with international benchmarks.

3 Linking Standardization and Interoperability

3.1 How Standardization Promotes Interoperability

Standardization plays a crucial role in promoting interoperability within the context of ERIGrid 2.0, which focuses on enhancing the European energy infrastructure, by establishing common communication protocols and interfaces that enable different devices, systems, and technologies to work together effectively. This is particularly essential for integrating diverse energy sources, such as renewable energy, into the grid. Additionally, by defining standardized data formats, the project ensures that data exchanged between various components of the energy system can be universally understood, facilitating better data sharing, analysis, and decision-making among stakeholders. Furthermore, standardization guarantees that equipment and technologies from different manufacturers can operate together without compatibility issues, which is crucial in a field like energy, where numerous technologies such as smart meters, energy storage systems, and grid management software, must interact seamlessly.

Moreover, establishing standardized testing and certification protocols allows for the reliable validation of technologies before their deployment, ensuring that all components meet specific performance and safety criteria critical for maintaining system reliability. Standardization efforts also lead to the development of best practice guidelines that provide recommendations for effectively implementing technologies and systems, promoting consistent approaches across various projects and regions, and thereby enhancing interoperability. By offering a clear framework for interoperability, standardization encourages innovation, allowing developers and researchers to focus on creating new technologies and solutions that align with established standards, thus minimizing the risk of incompatibility.

Collaboration among various stakeholders, including industry, academia, and regulators, is a hallmark of the standardization process, ensuring that the needs and perspectives of all parties are taken into account, which ultimately leads to standards that facilitate interoperability across the board. Lastly, standardization aligns with regulatory frameworks and policies, ensuring that interoperable solutions comply with legal requirements, a crucial aspect for the widespread adoption of new technologies in the energy sector.

3.2 Standards Enhancing Interoperability in PHIL-Testing

PHIL is a testing methodology that seamlessly integrates real hardware components with virtual models to simulate power systems. This approach is particularly valuable for developing and validating smart grid technologies and renewable energy applications. To ensure interoperability in PHIL and related domains, several standards have been established, each contributing to a structured and efficient testing environment.

Key standards supporting interoperability include IEC 61850, which defines a communication framework for substations, enabling seamless interaction between devices from different manufacturers. IEEE 1547 provides essential guidelines for the interconnection of DER with electric power systems, ensuring safe and reliable operation. IEC 61400-21 focuses on the electrical performance of wind turbines, while IEEE 2030 addresses smart grid interoperability, facilitating the integration of renewable energy sources and demand response technologies.

Additionally, IEC 62056 specifies protocols for data exchange in smart metering, enhancing communication between utilities and consumers. ISO/IEC 30141 outlines an architectural framework for the Internet of Things (IoT), fostering interoperability among IoT devices within energy systems. IEC 61850-90-5 provides specific guidelines for PHIL testing, ensuring effective interaction between simulations and hardware components. Other notable standards, such as IEEE P2030.5 for demand response technologies and DNP3 for control system communication, further enhance interoperability by defining common protocols for data exchange across devices from diverse manufacturers.

3.3 IEEE WG P2004 and Its Role in HIL Simulation

In the broader context of HIL-based testing, the IEEE WG P2004, officially titled the IEEE Recommended Practice on "Standard Hardware-in-the-Loop (HIL) Simulation-Based Testing of Electric Power Apparatus and Controls" and established comprehensive methodologies for evaluating electric power systems. The resulting standard IEEE Std 2004-2025 [14] offers essential recommendations for designing, deploying, and executing real-time HIL setups, ensuring rigorous validation of power apparatus and control mechanisms. It integrates analytical frameworks for stability, accuracy, and sensitivity assessments alongside experimental case studies to enhance reliability.

Although IEEE Std 2004-2025 was developed independently as a standard, it has benefited significantly from insights gained through research projects such as the sensitivity framework [13] elaborated within ERIGrid 2.0. The predecessor, ERIGrid, contributed to shaping international best practices in HIL simulation methodologies. Building on this foundation, ERIGrid 2.0 has further advanced the principles of IEEE Std 2004-2025, supporting its refinement through ongoing research, infrastructure improvements, and innovative testing approaches. By leveraging insights from both ERIGrid phases, IEEE Std 2004–2025 remains a crucial standard in the validation and optimization of energy system technologies, ensuring robust, high-performance solutions for modern power networks. The IEEE Std 2004-2025 is actively maintained by the IEEE Power Electronics Society and the Standards Committee PEL/SC, with contributions from leading experts. Designed to address the growing complexity of electric power systems, it plays a critical role in overcoming challenges related to real-time simulation fidelity and interoperability. Its structured approach enhances the reliability of HIL-based testing, making it an indispensable tool for researchers, engineers, and industry professionals engaged in advanced energy solutions.

3.4 Aligning Standards with Interoperability Goals

ERIGrid 2.0 achieved significant outcomes that effectively align standards with interoperability goals within the energy sector. By actively engaging a diverse range of stakeholders, including industry representatives, technology developers, regulatory bodies, and academic institutions, ERIGRID 2.0 ensured that the developed standards reflected the needs and perspectives of all parties, facilitating broader buy-in and adherence. Specific interoperability goals were established to address the challenges of integrating diverse technologies and systems, while existing international and national standards were leveraged to expedite the standardization process and promote consistency across projects.

Comprehensive interoperability frameworks were created, outlining necessary protocols, interfaces, and data formats for seamless communication between systems, and these frameworks are adaptable to future technological advancements. The project also documented and disseminated best practices for implementing interoperability standards, encouraging the development of modular technologies that can easily integrate with existing systems. Successful pilot projects tested these standards in real-world scenarios, providing valuable insights for improvement, while continuous feedback mechanisms were established to ensure that the standards remain relevant. Education and training programs were offered to raise awareness about the importance of interoperability, and monitoring and evaluation processes were set up to assess the impact of the standards on system performance. Finally, collaboration with recognized standardization organizations ensured that the developed standards are widely recognized and adopted, collectively contributing to a more integrated, reliable, and sustainable European energy system.

4 Summary and Recommendations

Standardization is a cornerstone of the energy sector and plays a critical role in ensuring that different components work together effectively. This is essential for maintaining safety and reliability, reducing costs, fostering innovation, and enabling global trade and cooperation. Without standardization, the interoperability between various energy components would be compromised, leading to inefficiencies and potential safety hazards. In the realm of smart energy systems, standardization becomes even more crucial. It supports effective communication and data sharing, ensuring that these systems can scale efficiently. Moreover, it contributes to cost efficiency and regulatory compliance, which are vital for the overall effectiveness and advancement of smart energy technologies.

National and international policy frameworks play an active role in shaping standardization efforts in the energy sector. Organizations like the European Union, OECD, ISO, IEC and IEEE are at the forefront, demonstrating a global commitment to establishing and implementing standards. These policies significantly influence how standards are developed, adopted, and implemented. Interoperability in energy systems is a key area where standardization is indispensable. By setting common communication protocols, data formats, and testing procedures, standardization ensures that diverse energy technologies and systems operate together seamlessly. IEEE P2004 stands as a vital standard within the realm of Hardware-in-the-Loop (HIL) Simulation, providing a robust framework for the testing and validation of power and energy systems. Its comprehensive methodologies and recommendations ensure the reliability and accuracy of evaluations. Ultimately, this standard serves as an indispensable resource for professionals in the energy sector, facilitating the development of innovative and high-performance solutions for modern power networks.

Initiatives like ERIGrid 2.0 play a significant role in aligning standards with interoperability goals. Through stakeholder engagement, the development of interoperability frameworks, pilot projects, and collaborations with standardization organizations, such projects contribute to creating a more integrated, reliable, and sustainable energy system. Ongoing efforts are essential to develop and harmonize standards in rapidly evolving areas such as artificial intelligence in grid management, advanced energy storage solutions, electric vehicle charging infrastructure, and the integration of hydrogen and other alternative energy carriers. These emerging technologies present unique challenges that require focused attention to ensure that standards keep pace with innovation.

As digitalization continues to transform energy systems, it is vital to enhance cybersecurity standards to safeguard critical infrastructure against evolving threats. Robust and adaptive cybersecurity measures will help protect these systems from potential vulnerabilities, ensuring their resilience in an increasingly connected world. Despite significant progress in the field, further international collaboration is necessary to promote global compatibility and facilitate the seamless integration of energy systems across various regions and markets. Addressing regional variations and needs within a unified global framework will be crucial for effective cooperation and standardization.

Moreover, advancing testing and validation methodologies is essential for ensuring the reliability and performance of increasingly complex energy systems. Building on initiatives like ERIGrid 2.0, additional research into methodologies such as HIL-based simulation and digital twins will play a key role in validating system performance and enhancing operational efficiency.

Lastly, developing comprehensive standards for data interoperability and sharing will be critical for leveraging the vast amount of data generated in modern energy systems. Establishing clear protocols and formats will facilitate effective communication and collaboration among diverse technologies, ultimately maximizing the potential of smart energy solutions.

References

1. Agency IRE (2013) International standardisation in the field of renewable energy. https://www.irena.org/-/media/Files/IRENA/Agency/Publication/2013/International_Standardisation_in_the_Field_of_Renewable_Energy.pdf. Accessed (Insert Date of Access Here)
2. Bates L (2025) Goodale: leveraging the global research infrastructure to characterize the impact of national science foundation research. https://www.researchgate.net/publication/387975427_Leveraging_the_Global_Research_Infrastructure_to_Characterize_the_Impact_of_National_Science_Foundation_Research. Accessed 03 Apr 2025
3. Commission E (2024) About the european research area (2024). https://european-research-area.ec.europa.eu/about-era. Accessed: 01 Oct 2024
4. Commission E (2024) European research infrastructure consortium (ERIC). https://research-and-innovation.ec.europa.eu/strategy/strategy-research-and-innovation/our-digital-future/european-research-infrastructures/eric_en. Accessed 01 Oct 2024
5. Commission E (2024) European research infrastructures. https://research-and-innovation.ec.europa.eu/strategy/strategy-research-and-innovation/our-digital-future/european-research-infrastructures_en. Accessed 01 Oct 2024
6. Commission E (2024) Funding and grants for horizon Europe research infrastructures. https://rea.ec.europa.eu/funding-and-grants/horizon-europe-research-infrastructures_en. Accessed 01 Oct 2024
7. Commission E (2024) Research infrastructures under horizon Europe—policy and strategy. https://research-and-innovation.ec.europa.eu/funding/funding-opportunities/funding-programmes-and-open-calls/horizon-europe/research-infrastructures_en#policy-and-strategy. Accessed 01 Oct 2024
8. Education D (2025) National research infrastructure and industry engagement. https://www.education.gov.au/download/13738/national-research-infrastructure-and-industry-engagement/26732/document/pdf. Accessed 03 Apr 2025
9. ESFRI: glossary (2025). https://www.esfri.eu/glossary. Accessed 03 Apr 2025
10. EUR-Lex: Legal content on research infrastructures (2024). https://eur-lex.europa.eu/legal-content/EN/TXT/?uri=LEGISSUM%3Ari0005. Accessed 01 Oct 2024

11. Group A (2025) Iso 50001: a guide to energy management and efficiency. https://www.arribatec.com/iso-50001-a-guide-to-energy-management-and-efficiency/. Accessed 03 Apr 2025
12. IEC: energy efficiency—iec international standards (2025). https://www.prd.iec.ch/taxonomy/term/796. Accessed 03 Apr 2025
13. Lauss G, Feng Z, Syed MH, Kontou A, Paola AD, Paspatis A, Kotsampopoulos P (2022) A framework for sensitivity analysis of real-time power hardware-in-the-loop (phil) systems. IEEE Access 10:101305–101318. https://doi.org/10.1109/ACCESS.2022.3206780
14. Lauss G, Lundstrom B et al (2025) (Members IEEE WG P2004): IEEE std 2004-2025: hardware-in-the-loop (hil) simulation based testing of electric power apparatus and controls (approved 12 Feb 2025). https://ieeexplore.ieee.org/servlet/opac?punumber=10806535
15. OECD: Infrastructure governance (2025). https://www.oecd.org/en/topics/infrastructure-governance.html. Accessed 03 Apr 2025
16. OECD: Research infrastructure (2025). https://www.oecd.org/en/topics/research-infrastructure.html. Accessed 03 Apr 2025
17. Research Infrastructures ESF (2024) Esfri—European strategy forum on research infrastructures (2024). https://www.esfri.eu/. Accessed 01 Apr 2024

Open Access This chapter is licensed under the terms of the Creative Commons Attribution 4.0 International License (http://creativecommons.org/licenses/by/4.0/), which permits use, sharing, adaptation, distribution and reproduction in any medium or format, as long as you give appropriate credit to the original author(s) and the source, provide a link to the Creative Commons license and indicate if changes were made.

The images or other third party material in this chapter are included in the chapter's Creative Commons license, unless indicated otherwise in a credit line to the material. If material is not included in the chapter's Creative Commons license and your intended use is not permitted by statutory regulation or exceeds the permitted use, you will need to obtain permission directly from the copyright holder.

Summary and Future Directions

T. I. Strasser, M. Calin, and L. E. Ramos Perez

Abstract This chapter summarizes the major findings presented throughout the book by recapping the ideas, conclusions, and future work described in each part. The content focuses on highlighting the complexity of testing and validating multi-energy systems and the developments that support this activity, resulting from the ERIGrid 2.0 project work.

1 Main Findings

The complexity of current energy systems requires the incorporation of specially designed methodologies to make testing and validation activities easier to carry out and replicate. The HTD approach provides a holistic framework for describing tests in such systems. The implementation of the developed ERIGrid 2.0 methodology allowed for a description of both the system structure and the experiment in a holistic manner (e.g., for the development of CPES applications). Beyond the usefulness of this approach, further improvements were developed as a form of *extensions* to extend its flexibility and applicability in other projects. A set of TCs was developed and later published in a dedicated open-access repository to provide an easily accessible reference on the use of the HTD extensions in further laboratory tests. Moreover, an approach which allows for an easy online publication of TCs as webpages by converting Word files into an online repository was developed through GitHub. For this purpose, a legacy post-project is ensured by keeping the website active for further developments.

T. I. Strasser (✉) · M. Calin
AIT Austrian Institute of Technology, Vienna, Austria
e-mail: thomas.strasser@ait.ac.at

M. Calin
e-mail: mihai.calin@ait.ac.at

L. E. Ramos Perez
European Distributed Energy Resources Laboratories (DERlab) e.V., Kassel, Germany
e-mail: leonard.ramos@der-lab.net

Another challenge on complex systems like CPES lies in the management of testing and validation and their reproducibility. ERIGrid 2.0 developed a systematic approach for assessing uncertainty in experiments together with benchmarks facilitating their design, documentation, and reproducibility. For this purpose, three new benchmarks were developed. The developed benchmarks serve as conceptual guides and reference models for representing and assessing capabilities of realistic systems involving multi-sector coupling, grid forming in microgrids, and power system-ICT interaction. Furthermore, the developed USAT tool provides support for the uncertainty analysis (which is crucial for experiment realization) and complements the improved HTD extensions and the validation methods in multi-energy systems. Moreover, these advancements are also promoting research and development in the field by providing useful tools.

Some applications of multi-energy systems co-simulation were also explored. Aforementioned multiple domains (specifically thermal and electrical) benchmark systems were modelled by using tools such as Mosaik, DisHeatLib and Python. Such simulations gave insights on issues associated with the initialization of the simulation (e.g., to reach a steady state or pattern) and alternatives to overcome them. Moreover, when implementing advanced real-time simulation, several requirements should be considered, including robust coupling interfaces for accurate modelling, rigorous synchronization and low latency.

Deepening into RTS and PHIL techniques for multi-energy systems, several challenges that are hindering its robust and widespread adoption were exposed. A narrow permissibility for time delays to keep system stabilities, as well as high systems sensitiveness to disturbances, arise as major issues. ERIGrid 2.0 achieved advancements in this last point, allowing to perform multi-energy systems experiments based on PHIL within a broader range of operational conditions (e.g., by introducing SP compensation or a novel VSI approach). Moreover, the developed framework for PHIL sensitivity analysis demonstrated its effectiveness for testing both elementary and complex power system components. Results from tests of advanced applications such as RTS/SIL, CHIL, PHIL and combined CHIL/PHIL demonstrated their usefulness for studying different dynamics in modern power systems. Further applications include the development and validation of DTs, testing of grid forming control, cyber resilience in CPS, etc.

Bringing the experimentation to a higher scale like with the interconnection of multiple RTIs geographically distributed. Here, several challenges arise, such as complexity for repeating or resuming experiments, signal interpretation alignment (e.g., differences in labels, units, scales, conventions, etc.), time delays, and clock synchronization, among others. To deal with such challenges, a series of tools focussing on communication harmonization, application of inter-communication, and configuration management can be implemented under an automated workflow. Further leverage of cloud computing technologies involved in multi-RI experiments becomes possible when implementing RIasC functionalities.

During the tests performed, the benchmark developed within ERGrid 2.0 simulated throughout software results in a suitable representation of multi-energy system dynamics as, for instance, for thermo-electrical systems with power-to-heat coupling. Moreover, during the demonstration of multi-energy system coupling for RTs geographically distributed, it was evidenced how ERIGrid 2.0 methodologies and tools for integrated testing, validation, and simulation can be used to support system integration in multiple domains.

By addressing key challenges related to power system stability, interoperability, and real-time control validation, the project showcases how user feedback has significantly influenced experimental setups and tool development. The collaboration between researchers and users led to improved data exchange tools and enhanced testing methodologies, ultimately ensuring that RTIs remain essential for driving innovation in smart energy systems technologies.

By offering a diverse array of resources, including open-source e-learning tools, virtual laboratories, and hands-on training opportunities, the project fosters both theoretical understanding and practical expertise. This holistic framework not only equips learners with the essential skills required to navigate the complexities of smart energy systems but also promotes interdisciplinary collaboration across key fields such as ICT, power systems, and control engineering. As the energy sector continues to evolve, the commitment to modern and accessible education within the ERIGrid 2.0 initiative ensures that the next generation is well prepared to meet the emerging industry challenges and contribute to sustainable solutions.

Standardization is essential in the energy sector as it ensures interoperability, safety, and reliability between various systems. This foundation promotes innovation, reduces costs, and enables effective global cooperation, particularly within smart energy technologies. In this context, the IEEE Std 2004–2025 standard plays a vital role, specifically within HIL-based experiments. It provides a robust framework for testing and validating electric power systems, ensuring the reliability and accuracy of evaluations. The comprehensive methodologies outlined in IEEE Std 2004-2025 support professionals in developing innovative, high-performance solutions for modern power networks, reinforcing the importance of standardization and policy in advancing the energy sector.

2 Outlook

Based on the outcomes and activities of ERIGrid 2.0 described previously, the future pathway for the European RTIs emphasizes enhanced interconnectivity and advanced simulation capabilities. This involves scaling up complex, geographically distributed multi-RI experiments, supported by refined tools for harmonization, automation, signal alignment, and configuration management, potentially leveraging cloud integration through RIasC concepts. Concurrently, RTIs are expected to advance their capabilities in complex RTS and PHIL testing, building on ERIGrid 2.0's improvements to handle broader operational conditions and system sensitivities while also improving multi-domain co-simulation fidelity and adhering to robust validation standards like IEEE Std 2004–2025 for HIL [1].

These enhanced RTI capabilities will underpin future work focused on critical emerging energy system challenges. This includes deeper investigation and validation of multi-energy systems and sector coupling, the extensive use of RTIs for developing and validating Digital Twins, creating focused testbeds for advanced grid-forming control strategies, and establishing methods to assess the cyber resilience of CPES. Directly stemming from ERIGrid 2.0's activities, future efforts will involve the continued evolution of methodologies like the extended HTD framework [2], the utilization and potential expansion of the project's benchmarks as standard references, and the refinement of tools for rigorous uncertainty assessment like USAT. Crucially, dissemination through maintained open-access resources (like TCs) and the sustained development of comprehensive education and training programs remain vital for capacity building, ensuring that the collaborative cycle of user feedback continues to drive innovation in smart energy system technologies.

The developed benchmarks [3–5] can be further improved for a greater modularity, towards a higher adaptability to other contexts. Thus, future work should focus on improving their compatibility with multiple simulation platforms and extend its use to other research environments.

References

1. Lauss G, Lundstrom B et al (2025) (Members IEEE WG P2004): IEEE std 2004–2025: hardware-in-the-loop (hil) simulation based testing of electric power apparatus and controls (approved 12 Feb 2025). https://ieeexplore.ieee.org/servlet/opac?punumber=10806535
2. Holistic test description templates, erigrid (2019). https://github.com/erigrid/holistic-test-description
3. Erigrid2/benchmark1 (2025). https://github.com/ERIGrid2/benchmark-model-electrical-network. Accessed 20 Apr 2025
4. Erigrid2/benchmark2 (2025). https://github.com/ERIGrid2/benchmark-model-multi-energy-networks. Accessed 20 Apr 2025
5. Erigrid2/benchmark3 (2025). https://github.com/ERIGrid2/benchmark-model-electrical-ict. Accessed 20 Apr 2025

Open Access This chapter is licensed under the terms of the Creative Commons Attribution 4.0 International License (http://creativecommons.org/licenses/by/4.0/), which permits use, sharing, adaptation, distribution and reproduction in any medium or format, as long as you give appropriate credit to the original author(s) and the source, provide a link to the Creative Commons license and indicate if changes were made.

The images or other third party material in this chapter are included in the chapter's Creative Commons license, unless indicated otherwise in a credit line to the material. If material is not included in the chapter's Creative Commons license and your intended use is not permitted by statutory regulation or exceeds the permitted use, you will need to obtain permission directly from the copyright holder.